The Krypto Economy
Andrea Bauer

Author/Editor: Andrea Bauer
D.DAY Co-Founders: Andrea Bauer, Boris Moshkovits
Lector: Allison Krupp
Graphic Design: Klute Agency/Diego Calvo

© 2017 Andrea Bauer
Publisher: tredition GmbH, Hamburg
ISBN (Paperback): 978-3-7439-5171-6
ISBN (Hardcover): 978-3-7439-5172-3
ISBN (E-Books): 978-3-7439-5173-0

Printed in Germany

Bibliographic information published by the National Library: The German National Library lists this publication in the Deutschen Nationalbibliografie. Detailed bibliographic data are available on the Internet at http://dnb.d-nb.de.

Index

Preface 8
Introduction 11
1. A Changing Narrative:
   From Hacking to Storytelling 16
2. What is the Colour of Money?
   The Myth of Economics 58
3. How Digital Activism is
   Changing the World 90
4. New Work Order: Who is
   in Charge? 134
5. Disruptive Powers and the
   Misfit Economy 174
6. Blockchain: Rewriting the
   Global Governance 216
7. The Brave New Art World 246
Thanks! 273

WHAT IS THE KRYPTO ECONOMY?

The term is derived from the idea of cryptography, the practice and technique of hiding digital information to allow a secure exchange. With this book, we extend the meaning, exploring the empowering and destructive forces of a hidden economy, which underlies the reality of the digital era. With Kryptonite—the fictional mineral that both nearly annihilates Superman and gives him incredible powers—acting as a basic narrative, we illustrate the digital age's radiating danger and also its undeniable force for good.

Preface

The technological evolution continues. Constantly an old world is dying, and a new one is born. Future ideas exist as seeds within the social fabric, awaiting expression by thinkers and creators.

Already, new technologies are changing our lives in extreme ways. They affect our human interactions, global understanding, business transactions and environmental exploitation. Technologies such as Artificial Intelligence, Blockchain, Internet of Things, Robotics, Autonomous Cars, Nanotechnology, Genomics, Biotechnology, Virtual Reality, Quantum Computing and 3D-Printing are not only driving new opportunities, but also pressuring our society to take a stand towards a more sustainable future. Perhaps most formidably, these technologies have the capacity to grow exponentially – taking us leaps and bounds toward a more utopian reality, one in which poverty and hunger are wretched memories. With the beauty of speed, we can use these exponentially growing technologies to create resilient cities and communities with responsible consumption in mind.

That said, this same technology has the undisputed power to catapult us into a postmodern surveillance dystopia, akin to novels of Philipp K. Dick[1] or George Orwell[2]. At the hands of people in power, these technologies can be the most dangerous facilitators of totalitarianism. What do we do in the face of this potential? We must understand. We must eliminate the risks. We must remain purpose-driven to build a responsible

1   Philip K. Dick was an American writer notable for publishing works off science fiction. He explored philosophical, social, and political themes in novels dominated by monopolistic corporations, authoritarian governments, and altered states of consciousness.

2   Eric Arthur Blair, with pen name George Orwell, was an English novelist and journalist. His work is marked by lucid prose, awareness of social injustice, opposition to totalitarianism, and outspoken support of democratic socialism.

future. We must lead the discourse to discover insights: with the potentials named and dangers stamped and tilted to see all sides. The Krypto Economy serves as an umbrella term on our journey to create this transparency with a sharp and critical eye, to piece through everything and articulate the problems constructively. As we do this, we can foster holistic, impact-driven technical solutions.

The discourse is the vehicle on our quest to observe, investigate, to finally gain in knowledge, and epiphanies. It is an attempt to characterize the truth, thus distinguishing it from falsity and illusions. Through non-institutional discussions, we aim for individual education, which will ultimately create a higher universal level of consciousness. This will empower people to join the conversation, providing them a basic set of questions and answers. The world becomes more informed. Inch-by-inch, person-by-person. We set out for the truth, actively forming our social economics in a digital age. It's the voice of the many that create the Zeitgeist[3]. And that Zeitgeist can decide upon a utopian, rather than dystopian future. Our economy isn't just an organizational framework for our societal, economical and political daily grind. It's a mirror on the current state of our consciousness. It's the perceived narrative of the human nature and its social interconnections. It reflects our values. And it encompasses official corporations, markets, and states, alongside criminal organizations, secret power networks, miracle healers, gangsters, or hackers. Everyone is part of the discourse. Everyone forms this economy.

A great era has begun. We are liberating ourselves. We are growing up and must take a responsible and courageous first step towards a utopian tech world. Heed our warning and listen to our appeal. Our future is up for grabs, but it's also at stake.

— Andrea Bauer

3   *The Zeitgeist, or the spirit of the age or time, is the dominant set of ideals and beliefs that motivate members of society in a particular period in time.*

Introduction

Over the past few years, we've begun to question the liberating and egalitarian vision of a digital era. With the revelations of Edward Snowden in 2013, it became clear that the Information Age would not provide us with a warm and fuzzy future, per se. Instead, we woke up in a surveillance and advertising apparatus. Too quickly, discourse ebbed away, along with the attention span of the public media, with little consequence.

With this in mind, Andrea Bauer and Boris Moshkovits began a think tank and discourse salon called D.DAY[4] to continue a critical, but constructive discourse to reflect on the future of our tech world; one that considered the current, rapidly evolving digital climate.

The first salon took place in 2014 at Soho House Berlin. Its aim was to

4   http://www.ddaynetwork.com

reflect on the realities and effects of the digital age, opting for the panel discussion format. Each time, they invited two protagonists of diverse special fields, who provided particular insights and were open to finding interconnections and phenomena. There were laughs, friendly bickering, surprising stories, but most importantly, honest discussions between people who comprehend their fields to the bone, and gave tentative solutions for future scenarios.

With the goal to create a holistic understanding of the effects of our digital age, the event brought techies, activists, artists, philosophers, authors, and misfit entrepreneurs together to link their experiences, argue, and laugh. The discussions brought comprehension to those not in the "know" in the tech world: steps to take, moves to make, and simple connections to consider. Over a series of seven events, the conversations touched on topics like mass surveillance, crypto currencies,

Blockchain, virtual art world, digital activism and the future of governance.

This book is a compendium of these seven panel discussions, which delved deep into the technical sphere, using laymen's terms to have an upbeat, informed discourse. With this book, you can fill your knowledge gaps, and join the discourse to become a creature of this vibrant world. And this "new tech order" can help you make decisions for a more responsible future. The future is ours to mold. And we have so much to accomplish.

ANDREA BAUER is an innovation designer, tech philosopher and author. She is the founder of BEAM Studio, an innovation firm that applies novel technologies and methods to create cutting-edge services, products, and business models. Always fascinated by the question of how technology can improve our everyday lives, she is pur-pose-driven, working to accelerate her efforts towards a more responsible future. She is also partner and co-founder of the business collective katapult:NOW and the event platform D.DAY Network, an international do- and think-tank of creative influencers, who aim for positive impact by reflecting and actively using the effects of a digital era.

BORIS MOSHKOVITS is a cultural entrepreneur, who has served as Head of Digital Media at Ringier Publishing in Germany, working to define the digital strategies for publications such as *Cicero* and *Monopol*. Prior to this, he was the founding director of the Jewish Museum and Tolerance Center in Moscow, which won the Award for Best Museum in Russia in 2012 due to its cutting-edge digital and interactive presentations and exhibition. As the founder and publisher of *Berliner Magazine* in 2002, Boris was part of the first wave of magazines that was instrumental in bringing attention to the "berliner attitude" and lifestyle in other metropolises, like New York, London and Paris. In 2014, he co-founded D.DAY Network with Andrea Bauer.

1
A Changing Narrative: From Hacking to Storytelling

Angela Richter
Jacob Appelbaum

PANEL DISCUSSION, BERLIN,
MARCH 15, 2014

Who owns the future of the Internet?
Disclosures of mass surveillance and a
fast-growing epidemic of cyber crime
have corrupted the "utopia" of a free
and equal Internet. Alongside this, in the
last 10 years, the Internet has become
crucial to everyday life, enabling people
to interact in personal and professional
ways and build connections and
businesses.

But major companies such as Facebook,
Google and Apple have created predominant services and communities—
whose free use we pay for with our data.
And while activists invoke digital civil
rights through encryption, hacking and
campaigning, governments continue
to violate human rights in the name of
fighting global terrorism.

We can discuss the ownership of the
future structure of the Internet, much like
we discuss the organization of established

infrastructures such as energy and transportation. Many questions arise: What can we learn from Snowden and Assange? How can we be free in a digital world? What happens to our privacy? Does governmental security inevitably mean total control?

The impact of mass surveillance and the uncontrolled use of big data have inspired cultural producers to change the narrative, working to regain privacy and dignity for everyone. In light of this, D.DAY will host leading minds from digital activists to theatre directors, to learn more about their narratives of the Internet in a post-Snowden world. To transform the use of the Internet from a commercial to a social space, we will focus on the role of transparency and personal dignity.

PANELLISTS

There are no greater people to alter the narrative of privacy and dignity than Angela Richter and Jacob Appelbaum. Their careers mark the path from hacking to storytelling in this post-Snowden world. Both have had first-hand experience with Julian Assange and Edward Snowden, two of the most-wanted and most-secluded men of our times. And their stories propel incredible, if at-times terrifying, insights.

The collaboration between Jacob and Angela marks a cultural shift in raising awareness for the growing loss of privacy. We work to analyse their brand of storytelling and how this brings awareness to digital mass surveillance. It lends reflected insights and creates immersive experiences. And in the end, it gives their audience a voice.

ANGELA RICHTER is a playwright, theatre director, and activist. In the years leading up to this discussion, she focused her creative and activist strengths on mass surveillance and transparency. Her on-going interviews with the WikiLeaks founder, Julian Assange, led to the play *Assassinate Assange*, in which she examines various aspects of Julian's persona and the shift in the public's perception. In the second part of the series, *Assassinate Assange Reloaded*, Jacob Appelbaum performed alongside her. Since then, the director not only visited the WikiLeaks founder in London regularly, but has engaged in panel discussions and with various print media, including *Der Spiegel*, *Monopoly*, *Interview Magazine*, *Review* and *Next Society*, speaking for Internet activists and hacktivists. In the season 2014/15, Richter scheduled a large-scale project entitled *Die Avantgarde der Supernerds* in co-production with the German TV WDR, dealing with the life and work of digital dissidents.

JACOB APPELBAUM is an American independent journalist, computer security researcher, artist, and hacker. He has been employed by the University of Washington, and was a core member of the Tor project, a free software network designed to provide online anonymity. Appelbaum is also known for represent-ing WikiLeaks. He has displayed his art in a number of institutions across the world and has collaborated with artists such as Laura Poitras, Trevor Paglen, and Ai Weiwei. His journalistic work has resulted in a number of books, as well as publication in *Der Spiegel* and elsewhere. Appelbaum has repeatedly been targeted by U.S. law enforcement agencies, which obtained a court order for his Twitter account data, detained him at the U.S. border after trips abroad, and seized his laptop and several mobile phones.

Angela, in your play *Assassinate Assange* you say: "Die Zukunft gehört den Nerds," or "The future belongs to the nerds." Do you really believe that the hackers and the nerds are the new avant-garde?

ANGELA: This is a very provocative sentence, especially in the context of art and theatre, because we have a long tradition of being the avant-garde, an art format that always influenced society. It's a little bit radical and also meant to be a bit provocative, but I think it is true. Even more than I would have liked it to be.

We see it in personas like Julian Assange[5] or Edward Snowden[6] and the very fast paradigm shift they achieved with their actions. Things changed so fast. I came up with this "the future belongs to the nerds" statement almost two years ago and put it on stage. At that time, I couldn't foresee what would happen today.

Jacob, you started out as what we call a "hacker" or "hacktivist," focusing on encryption, but now you work with Angela Richter on plays, with Laura Poitras on videos, and as a journalist for *Der Spiegel*. Why did you choose these more artistic ways to express yourself?

JACOB: In some cases, there are very serious legal consequences when dealing with topics of secret intelligence, unless you are a special type of person. Being a journalist allows you to talk about topics in the public interest with some privileges you might otherwise not have.

5   *Julian Paul Assange is an Australian computer programmer, publisher and journalist. He is editor-in-chief of the organization WikiLeaks, which he founded in 2006.*

6   *Edward Joseph Snowden is an American computer professional, former Central Intelligence Agency (CIA) employee, and former contractor for the United States government who copied and leaked classified information from the National Security Agency (NSA) in 2013. His disclosures revealed numerous global surveillance programs, many run by the NSA and the Five Eyes Intelligence Alliance with the cooperation of telecommunication companies and European governments.*

My political voice as a programmer is silenced if I want to talk about some of the issues that we face, like the issues of mass surveillance, unless I am a credentialed journalist. I have worked with *Der Spiegel* for many years, probably since 2010.

I thought it was very important to continue that work, especially with some of the topics that Edward Snowden has released to the public. It was important to try to change the dialogue a lot. For many years, people thought that myself and other people talking about mass surveillance were just crazy. The joke is on you! Unfortunately, we weren't crazy enough.

> It's important to win the culture war. <

I think it's important to win the culture war, so to speak, or the culture conflicts that are happening right now. To do that, you have to involve the art world. You have to write. You have to take every possible angle to introduce it into people's lives to make it relevant, so that people can begin to contextualize and understand what is actually happening in a way that is meaningful to them.

Angela, a long time ago, you produced *Ödipus* at Salzburger Festspiele. I think this was the moment when you realized that you are not very interested in classical plays, but more into realistic plays. What kind of narratives do you prefer in theatre? What is the role of the theatre today?

ANGELA: I became really bored, and then I came to realize: when I'm bored at my own rehearsals, what should I expect from the audience? It is not that I don't like classical plays. I admire Shakespeare. I love to read it, but I just don't have the feeling that I should put it on stage, because I always had the feeling that I'm building a museum. It became really boring to me, and I thought, "What else can theatre do?"

When I directed *Ödipus* in Salzburg, I learned about WikiLeaks[7], and I thought it was a really overwhelming idea to upload documents from secret sources and publish them on a website. It was just fantastic. I thought, "Why didn't somebody come up with the idea earlier?" It seemed too obvious.

At some point, I decided to only do research-based plays. I had done it once before, with the forbidden novel by Maxim Biller, *Esra*. With *Esra* I made a play about the whole legal case and a little bit about the rose war between them. The play *Assassinate Assange* was just the next logical step, since I was so interested in WikiLeaks.

Jacob, you already mentioned that you worked with *Der Spiegel*. What is the role of mass media and the press today when it comes to creating transparency and understanding topics like mass surveillance? Did media fail in reporting about mass surveillance? And how does investigative journalism help?

JACOB: Well, I think it depends. There are many different media outlets around the world. Right now, Germany is the freest place that we can publish about the NSA surveillance[8]. I cannot work as a journalist in the United States. Like in the United Kingdom, there is a Terrorism Case[9] as well as an Official Secrets Act Case[10]. People

7   *WikiLeaks is an international non-profit journalistic organization that publishes secret information, news leaks, and classified media from anonymous sources.*

8   *PRISM is a surveillance program under which the United States National Security Agency (NSA) collects Internet communications from at least nine major U.S. Internet companies. Its existence was leaked by NSA contractor Edward Snowden, who warned that the extent of mass data collection was far greater than the public knew and included what he characterized as "dangerous" and "criminal" activities.*

9   *The USA Patriot Act is an Act of Congress that was signed into law by President George W. Bush on October 26, 2001. With its ten-letter abbreviation expanded, the full title is "Uniting and Strengthening America by Providing Appropriate Tools Required to Intercept and Obstruct Terrorism Act of 2001".*

10   *Under The Official Secret Act or the U.S. Espionage Act, governments can prosecute journalism that undermines national security.*

like Sarah Harrison[11] cannot return to the United Kingdom as a result of those cases, and certainly not to continue to work and publish about these revelations.

So did the mass media fail? No. Actually, society is failing the media and the free press and the promise to a free press. And right now, Germany is one of the better places to work on these issues. It kind of makes sense, because these issues have almost nothing to do with German politics, or, let's say, Angela Merkel's mistakes. It is almost always the case that you're freer to criticize things from another system, when you exist almost entirely in a different framework.

> ## Did the mass media fail? No. Actually, society is failing the media. <

There are some exceptions. For example, Germany is particularly victimized. I revealed, for example, the Merkel scandal, when the NSA spied on Angela Merkel[12]. That was my story. I brought it to *Der Spiegel*, but it would not have been possible in the United States. That is not a failure of the media. That is a failure of the U.S. government in controlling its spy apparatus. Its protection of the press is just abysmal.

Is the press aware of the U.S. surveillance, if you compare the German and the U.S. press, for instance?

JACOB: The U.S. is, in theory, a very free country. I grew up in the United States, and I'm really proud of a lot of things that I experienced when I was there. But I also

11   Sarah Harrison is a British journalist, legal researcher, and WikiLeaks section editor. She works with the WikiLeaks Legal Defense and is Julian Assange's closest adviser. Harrison accompanied National Security Agency whistleblower Edward Snowden on a high profile flight from Hong Kong to Moscow while he was sought by the United States government.

12   NSA tapped German Chancellery for decades, WikiLeaks claims.

think nationalism is a sickness. I think that the U.S. is seriously ill with nationalism.

We have a theoretical freedom of the press, which states the press won't be regulated. But we are talking about people like Glenn Greenwald[13] in the public sphere, versus the Attorney General being questioned directly whether or not he will be prosecuted for writing about Edward Snowden[14]. The Attorney General essentially sidesteps this.

> The U.S. is seriously ill with nationalism. <

It is clear on paper that Glenn Greenwald is not a criminal. He is a journalist telling us things in the public interest. The Attorney General, when directly confronted, could not say that. He even said that people who were advocating for Edward Snowden, who were protected under The First Amendment[15], would be treated differently than legitimate journalists. Who is to say if people like Julian Assange, Laura Poitras[16], Glenn Greenwald, and myself are legitimate journalists?

13   Glenn Edward Greenwald is an American lawyer, journalist, speaker and author. He is best known for his role in a series of reports in The Guardian newspaper on the classified information made public by whistleblower Edward Snowden.

14   In November 2013, Attorney General Eric Holder indicated that the U.S. Justice Department is not planning to prosecute former Guardian reporter Glenn Greenwald, one of the journalists who received documents from Snowden and has written a series of articles based on the leaked material.

15   The First Amendment is one of the 10 amendments that constitute the United States Bill of Rights. The First Amendment to the United States Constitution prohibits the making of any law respecting an establishment of religion, impeding the free exercise of religion, abridging the freedom of speech, infringing on the freedom of the press, interfering with the right to peaceably assemble, or prohibiting the petitioning for a governmental redress of grievances.

16   Laura Poitras is an American director and producer of documentary films. She has received numerous awards for her work, including the 2015 Academy Award for Best Documentary Feature for Citizenfour, about Edward Snowden.

A Changing Narrative: From Hacking to Storytelling

Obviously, if we dare to advocate that we think Edward Snowden is a whistleblower, then we will be dealt with in a harsher manner. In a sense, and on paper, there is way more free speech in the United States than in Germany. In practice, I'm pretty sure I'd be in prison right now if I published the things I published with *Der Spiegel* in the United States, but in Germany, I'm free.

On paper, the German free speech is significantly more limited. The reality is that Julian Assange is stuck in this Ecuadorian Embassy as a result of the pressure of the United States government. Edward Snowden is stuck in Russia because of the pressure of the United States government. We're not really living up to those promises of free press.

Angela, you chose the number one enemy of the United States as a main character of your play and were looking at all the different perceptions of him—as a sexual harasser, as a Robin Hood of the Internet, as a saviour. What was the strongest reaction you got from audiences and media, and do you feel you are caught in the middle of a propaganda war?

ANGELA: It's been almost two years since we had the opening in Hamburg. Then, we went to Vienna, then to Berlin, and two months ago, we finally played in Cologne. The strongest reaction was actually when we showed the play in Vienna. The discussion around the rape case escalated, and for the first time in my life, I was really threatened by people. There were radical feminist groups, who wrote an open letter to the theatre to ban the play. I was extremely surprised to receive a reaction to seriously ban art from left people. They didn't even see the play. The reaction was only based on an interview I gave to *Der Freitag*. In the interview, they asked me if I think that Julian Assange is a rapist, and I said, "No," and elaborated on it. But they only picked up the "no" without the elaboration.

>For the first time in my life, I was really threatened by people.<

I was accused to be a supporter of rape culture, which, of course, I'm not. They also attacked the theatre and wrote: "No stage for rapists" on the outside walls. At that time, it was really difficult. We had a lot of discussion after every play with the audience. Most of the time, it was about the sexual things, not about the political aspects. I was attacked a lot. Some people were saying I'm just a groupie. It was ridiculous. Others sent threatening emails to the theatre.

I never experienced anything like this before, but I thought, okay, it's art. You can watch it. You can be against it. You can criticize it. But this aggressive reaction was intimidating. It would have been ridiculous for a theatre to forbid a play. I mean, we live in different times now, right?

Jacob, the technical and political world has become insanely complex. How would you describe the role of hackers and whistleblowers today?

JACOB: Interestingly, there is not a word for whistleblowing in German, as I understand it, which is kind of strange, since you guys have a word for everything. So, get on that, linguist!

It seems to me like the role of a whistleblower is actually the role of showing us the internal workings, to provide the transparency that is often missing from some of these structures. Julian Assange's big accomplishment is to recognize it when an agency—whether it's a government agency or a corporation or even individuals—puts forward one face to the world with a press release that only tells you what they want you to know. But if you have an internal document, you know what they really think.

It tells you the internal processes by which they decide things. A whistleblower can bring out the truth in how they actually act and how they really behave.

So a part of the role of a whistleblower in modern society is to be able to change the dialogue from the press release to the actual internal workings. Like what Daniel Ellsberg did when he was working with McNamara to leak the Pentagon Papers[17]. He showed that there is the official narrative, and then, there is the truth. In the truth, they admit to lying, they admit to killing people, they admit to all these things they would never say in a press release because they are not proud of it. That allows us to actual have a functioning democracy, when it comes to the validity of these institutions and the decisions they make. If we wish to understand how some of these organizations actually function, and thus to see if they are valid, we must have leaks. We must have the unvarnished truth.

Of course, there is a moral or social tension to it, because in theory, you should be able to have an organization without worrying about whistleblowers and always being part of the public dialogue. But some organizations just go too far. For example, if an organization supports death squads[18], maybe we should reveal that? That is clearly in the public interest. Or perhaps they have mass surveillance of the entire world. Hey, what

17 The Pentagon Papers, officially titled United States – Vietnam Relations, 1945–1967: A Study Prepared by the Department of Defense, is a United States Department of Defense history of the United States' political-military involvement in Vietnam from 1945 to 1967. The papers were released by Daniel Ellsberg. He had worked on the study, and first brought it to the attention of the public on the front page of The New York Times in 1971. A 1996 article in The New York Times said that the Pentagon Papers had demonstrated, among other things, that the Johnson Administration "systematically lied, not only to the public but also to Congress."

18 A death squad is an armed group that conducts extrajudicial killings or forced disappearances of persons for the purposes of political repression, genocide, or revolutionary terror.

about that? Some things are just over the top and blatantly illegal. In these cases, whistleblowing is clearly a way to go. It is not always that obvious, when it's a minor affair. But in the case of Edward Snowden and Chelsea Manning[19], it couldn't possibly be more obvious.

The role of the whistleblower is to act as entropy and to introduce some entropy into the overall discussion.[20] Even if the entropy is unpredictable, in many cases, it is truthful because it is truth written by the hand of the oppressors themselves. That is very powerful, because it means one single person can take an action that will discredit all of the most credible people, and show that, in fact, they are not credible at all. In truth, they just have a structure that supports them and makes them seem more credible.

That is a very important role, but it is also a very controversial one. Imagine if Snowden had been a Chinese citizen, for example. He probably would have gotten asylum in Germany, which tells us something about the way we really value whistleblowers. I think that is a little sad.

In the U.S., it is much the same. If someone from China had done what Snowden had done, he would have gotten asylum in the United States. There is a lot of political stuff wrapped up in that act of whistleblowing and how it is framed. Is it treasonous or being heroic?

There are also some questions on what makes a person a whistleblower and not a spy. I think acting in the public interest is clearly a defining element that makes a person

19   *Chelsea Elizabeth Manning (born Bradley Edward Manning) is a United States Army soldier who was convicted by court-martial in July 2013 of violations of the Espionage Act and other offenses, after disclosing to WikiLeaks nearly three-quarters of a million classified or unclassified but sensitive military and diplomatic documents. Manning was sentenced in August 2013 to 35 years imprisonment.*

20   *Entropy, in the sense of communication technology, has the meaning of a measure of the loss of information in a transmitted signal or message.*

a whistleblower, while selling secrets makes them a spy. That would be espionage. Telling the whole world that there is a criminal activity going on in their organisation is not what spies do. Spies don't blow their own operation. That doesn't make any sense to them.

Daniel Ellsberg originally said, "Courage is contagious." Julian Assange and Sarah Harrison quoted that statement. Is it true? Is courage contagious? Do you think that there are more and more whistleblowers joining to change the dialogue?

JACOB: I totally agree. I think courage is contagious, but courage is not the absence of fear. I think it is important to note that every person that we talked about has a lot of *Angst*—as is the German term for fear—about a lot of these issues, about making the right decision and really trying to change this dialogue.

> ## >Courage is contagious, but courage is not the absence of fear.<

There is courage, but it isn't without serious consideration, without serious side effects. Sarah Harrison can probably never set foot in her own country again. So being courageous comes with a lot of other things and should be thought of in a humble way. And I think, when Sarah is quoting that statement, it's meant in a humble way.

Talking about contagious, Angela, with your play, do you think it's somehow a call to action? When I watched your play, I feel you are torn between what is legal, what is illegal, what is true, what is not true. You are playing with different aspects, leaving the viewer to make his own mind. Do you want to activate them? If yes, how does activation work nowadays?

ANGELA: Yes. Actually, I want people to stop just thinking about it and feel activated. I don't feel that my play is taking a stand, so much. I myself, as a person, have developed a stand now, since I learned so much from people like Jacob and others from the scene.

Actually, Jacob and I met when all this was happening in Vienna. He wrote me an email and asked, "Can I help? I could sit on the panel and take on the discussion." The next time, he did it again at the HAU[21] in Berlin. With the discussions after the play, I wanted to activate the people because the play raised so many questions. And even the play has some provocations in it; it doesn't gives you a clear answer or takes an extreme stand. There is also humour in it. Sometimes, I even make jokes at Julian's expense.

As you saw in the play, I don't choose to have a guy with a white wig coming on stage. I decided to use 13 masked white gorillas, which, at some point, take off the masks and say something. I wanted to avoid putting an actor on stage to represent Julian Assange. That can appear ridiculous, especially if the character you are portraying is not a historical but a living person.

For me, it was an ideal setting, because many of the actors and actresses could have the role of Julian Assange at some point, like everybody could be him. We always had such a strong reaction from the audience in the end. This kind of reaction was something I could have only wished for when I did The Cherry Orchard by Chekhov[22] years ago, which is a really beautiful play. Afterwards, people just went home and thought "my world is okay," while actually, it is not! That is why I feel obliged as an

21   HAU means the Hebbel am Ufer and is a theatre and international performance center based in Berlin, Germany.

22   The Cherry Orchard is the last play by Russian playwright Anton Chekhov. It opened at the Moscow Art Theatre on 17 January 1904 in a production directed by Constantin Stanislavski. Although Chekhov intended it as a comedy, and it does contain some elements of farce, Stanislavski insisted on directing the play as a tragedy. Since this initial production, directors have had to contend with the dual nature of the play.

artist and a theatre person to shout out loud: "No, it's not okay".

It's very hard to activate people. Sometimes you really need to shock them, like with your current play *Überleben unter Überwachung* or *Survival under Surveillance*. In his role, Jacob got seriously injured on stage.

> ANGELA: I'm not so sure. I don't receive our actions as shocking. Maybe I have different standards. But Jacob can explain a little bit about the play, because it was very fragile and personal.

> JACOB: Originally, the play was supposed to be just a discussion on my experiences on surveillance with my long-time partner.

> My long-term partner and I split up long time ago. For various reasons, the theatre communicated with her about the play, and it turned from a panel into a play. Then, we had a four-act play along with a public lecture portion. She had written a huge amount of material for the intermediate points between the talk and the closure. It was emotionally and physically traumatic, as well, because we played the roles in a way that showed the internal mental states of our relationship. I don't know why we wanted to do this.

> Originally, we just wanted to talk about mass surveillance. Then, we were purging the things we had lived through in our relationship as a result of surveillance. It was extremely emotionally draining. We hired some real security guards to re-enact some of the experiences we had, like my disappearances. A number of times when we travelled together, I would go through a checkpoint and some big goons would take me away and tell my ex-girlfriend that I don't exist. She would ask, "Where is my boyfriend?" and they would say, "There is no boyfriend. What are you talking about?" That happened to us a bunch of times.

> She would ask,
"Where is my boyfriend?"
and they would say, "There is
no boyfriend. What are you
talking about?" <

On stage, it was a little rougher. They stripped me naked and threw me on the ground a little too hard, slashing my entire foot open. Then, they dragged me bleeding, kicking, and screaming around the stage, so there's a bloody drag mark.

Then, my ex-girlfriend talked about what it was like to be a woman living under surveillance and what it's like to meet and date people under surveillance. In a sense, she sort of described Eugene Ionesco's play *Rhinoceros*. [23] She described the modern female version of the madness of trying to maintain your integrity and your mental state in a surveillance state, where your partner is being targeted. You're also being targeted because of this notion of "valid targeted surveillance."

At the end of the play, I'm standing naked on stage without out glasses talking about the internal state. It was horrible, for sure. But at the same time, it was really cathartic. I think we were able to make it work really well, showing truly how we felt about living under surveillance, being a so-called "valid target" and the ways in which we were literally taken away or held or questioned. We showed it's corrosive, how it essentially eats away your soul.

23   *Rhinoceros is a play by Eugène Ionesco, written in 1959. Over the course of three acts, the inhabitants of a provincial French town turn into rhinoceroses; ultimately the only human who does not succumb to this mass metamorphosis is the central character, Bérenger, a flustered everyman figure who is initially criticized in the play for his drinking, tardiness, and slovenly lifestyle and then, later, for his increasing paranoia and obsession with the rhinoceroses.*

At the end of the play I was grabbed again. This time, I was already naked so they didn't strip my clothes, but they dragged me again and threw me outside of the theatre. This was very funny for the people standing outside of the theatre because all of a sudden, there was this strange, naked guy.

Then the lights go off, everyone starts to applaud and the lights go back on. Then, for security reasons, everyone who wanted to applaud us and thank us for the play was immediately removed from the theatre.

We wanted to show them the emotional state, how you have an emotional desire that is denied. We wanted to make them feel that arbitrarily and capricious feeling, which made them really angry in the end.

ANGELA: Yes, the audience was angry. They couldn't believe they had to leave. So many people just said, "No, we refuse to go." There was a guy from the theatre fighting with the audience for fifteen minutes because they insisted on the applause, but we wouldn't give it to them. That might have been the most provocative moment in the play: to break the rules, and say, "No, we will not give you that relief." It was very intense, and it was also very private.

I think it was a very good example on how private issues can became political. Normally, I do not do this fragile, very personal stuff on stage. But in this case, I thought the volume needed to be turned up to make the experience more intense for people to see what it means: the illusion of total freedom. Because suddenly, things can happen that affect our lives or the lives of our children.

What I found most shocking about this whole surveillance revelation is how the military, the intelligence services, and states are merging into this weird transnational thing. It's very frightening. I don't have the feeling that any of our politicians really has control over it. It is dynamic. It is covering nearly the whole world, and I don't have the feeling it's really controllable, not even by Obama himself.

JACOB: The whole surveillance situation was also really, intensely personal for my former long-term partner. Once, she woke up in the middle of the night with two men watching her sleep with night vision goggles. She looked up, and it was like some scene out of the movie *Silence of the Lambs*.

> You would call the police, unless the state itself is doing it to you. <

Naturally, in that case, you would pick up the phone and call the police, unless the state itself is doing it to you, which was exactly the case. So when she did call the police the next day, they laughed at her and refused to take a police report. It took three times and eventually getting the American Civil Liberties Union involved. Only then would the police take it seriously.

Part of this play was to talk about how this is the last outlet for her to deal with these things, because all of the normal channels were either silenced or lacking real purpose. That was very cathartic for her. It was very cathartic for all of us, but it was also very traumatizing to relive these things. I haven't actually been able to walk correctly since this happened last Saturday, because they actually hurt me so badly onstage. It partially ripped my tendon and my blood was everywhere. It was very "German art world."

Do you feel that there's a country out there that is not repressive, not restrictive, and listing you as a person of interest? Is there anywhere you do feel safe?

A Changing Narrative: From Hacking to Storytelling

JACOB: All states have problems. They are states. I tend to think that every place is what you make it, and right now, I feel relatively okay in Germany.

Most countries have intelligence services. Iceland is an example of a country that doesn't, but it has plenty of other problems. Julian Assange once pointed out that it has 100% literacy because it's xenophobic and racist. Basically, they're only letting in white people that are well educated.

I don't totally agree with him about that. It depends on what you look at, how you see it. I have certain privileges that are afforded to me. If I was a Muslim living on Oranienplatz in Berlin, I would probably not feel so great about Germany, for example. Every place is different.

For the secret service stuff, however, the U.S. is totally off the rails. It's really unbelievable. And Germany is also, to some extent, because the U.S. secret service has great influence over the BND[24] and the Verfassungsschutz[25]. Even some of the great countries that are, in theory, changing us, like Ecuador or Iceland, have problems as well.

No place is perfect. We should really be trying to talk about what we would like to see out of these states and try to make them better. In Germany, there is a parliamentary investigation about spying. And Germany can make a big difference in choosing a different paradigm. Hopefully that will happen. But if it doesn't happen with Germany, I'm not sure which country will actually start to make those choices soon. It is not the U.S., that's for sure.

24   *The Bundesnachrichtendienst or The Federal Intelligence Service is the foreign intelligence agency of Germany, directly subordinated to the Chancellor's Office.*

25   *The Bundesamt für Verfassungsschutz (BfV) is the domestic intelligence service of the Federal Republic of Germany.*

Have you been invited by Parliament or any official institution here in Germany to share your insights and your experiences? They always talk about Edward Snowden, but he is far away. You are here.

JACOB: Yes, the Council of Europe[26] and the European Parliament[27] both invited me to testify about the mass surveillance issues. I've done that. I was once invited to a CDU[28] event after the election, where they wanted me to come and talk about all mass surveillance issues. I said, "I really don't know anything about your politics, but I want to talk to you about the threat that this is to a free society." They gave me the date; they gave me the time. Then, the day before the event, they apologized and said that normally, they don't have members from civil society involved in their events. They had to uninvite me.

I think there's something interesting there. They recognized that I had something to contribute, but thought I wouldn't tow the line on all their other issues, so they didn't want me.

But Stroebele's[29] office was really fantastic. Stroebele asked me to explain some of the technical issues, and I explained what we have written about in *Der Spiegel* in my really bad, broken German. He was very thankful, and he told me that if I need some help, I should ask him.

26   The Council of Europe (CoE; French: Conseil de l'Europe) is an international organization focused on promoting human rights, democracy and the rule of law in Europe.

27   The European Parliament (EP) is the directly elected parliamentary institution of the European Union (EU). Together with the Council of the European Union (the Council) and the European Commission, it exercises the legislative function of the EU.

28   The Christian Democratic Union of Germany is a Christian democratic, and liberal-conservative political party in Germany. The leader of the party, Angela Merkel, is the current Chancellor of Germany.

29   Hans-Christian Ströbele, (born 7 June 1939) is a German politician and lawyer. He is a member of the German Green party.

A Changing Narrative: From Hacking to Storytelling

But realistically, I have the feeling that in terms of this issue, the German government doesn't really want to dig in. For example, read the work by the journalist John Goetz, a really fantastic journalist in Germany at NDR and Sueddeutsche Zeitung. He has been working a great deal on drone-related stuff and he has a project that is called *Geheimer Krieg*. He wrote a book and made a film.

Basically, he discusses Germany's role in the U.S. drone strikes. How the targeting happens, how the murder happens. These are assassinations that the German government participates in. And the German government is not looking into it. They are not inviting John Goetz to the parliament. The discussion is not actually as open as one would expect it to be, which is a little sad.

When you look at the new building of the BND, you are probably not very excited about the size, scope, and possibilities?

JACOB: It's actually incredible that the BND is getting a new building. It's amazing how large it is. The Secret Service has fundamentally undermined the notion of transparent, openness, and democratic processes. These things happen in secret. They use huge amount of mass data, and they use them for things that are not at all consistent with what we would consider "reasonable" in society. They use them for blackmail, for economic espionage, for assassinations. When we build buildings for people like that, we are supporting those kinds of activities, economically, literally, and politically, and that, to me, is terrifying.

I just can't believe that Germany is continuing down that path after a dictatorship.

People will bring up this notion of open societies and their enemies, saying you have to play this game to have an open society. You have to have some of these "city eaters," where they do the dirty deeds that the rest of society doesn't need to do. They are essentially making a commitment to anti-democratic ideals to maintain us as a democracy.

I think that is really a dangerous paradigm. For example, look at Norway and Breivik[30]. This guy murdered huge numbers of Norwegian people, but they did not do what the United States did after 9/11. They did not destroy their society to save it. Instead, they punished one person, using their specific laws to put this guy in jail. Sure, it won't bring those people back, but all the vengeance in the world that we see after September 11th won't bring those people back, either.

The difference is that the U.S. celebrates the deaths of people. When we assassinated Osama Bin Laden, as an example, there were people in the United States dancing in the streets.

>I just can't believe that Germany is continuing down this path after a dictatorship.<

When September 11th happened more than 10 years ago, the most horrifying thing to me and to most Americans was that people were dancing in the streets in support of some innocent or guilty people dying. We make that shift when we choose Secret Services over everything. We make a similar shift when we walk away from the basic Grundgesetzt[31].

Right now, these intelligent buildings are in the middle of our society, the cities. They're not hidden away. What is democracy, then? Is it an empty ritual?

30   Anders Behring Breivik is a Norwegian far-right terrorist who committed the 2011 Norway attacks.

31   The Basic Law for the Federal Republic of Germany (German: Grundgesetz für die Bundesrepublik Deutschland) is the constitutional law of the Federal Republic of Germany.

ANGELA: What I find very astounding is how big these organizations became. When I produced *Assassinate Assange Reloaded*, Edward Snowden had just revealed the NSA files. I met Julian, and we talked about that. He had a very interesting analogy for the situation.

He compared the surveillance system and the intelligence agencies with a liver, which had become too big in a body. The liver is supposed to detox and help you live. To a certain degree, every state needs intelligence for obvious reasons. It would be naïve to abandon it.

But when you get liver cancer, it gets out of hand. It takes too much energy and money. It gets out of proportion. The liver is killing the host that it lives on. At a certain moment, the system can collapse.

But Julian said that this is the optimistic version. The pessimistic version is that it doesn't collapse. It just goes on and on and on. That is actually more frightening.

JACOB: I'll give you a very concrete example of why I'm disgusted by the intelligence services. First of all, there's this notion that we need them. Fuck that. We don't need them. That is a total lie.

Some intelligent services have very specific jobs, which we do need. For example, stopping people from smuggling atomic weapons. We all agree that that should be done, but it doesn't need to be done in secret. We don't need to destroy our fundamental liberties to make that job happen. There are lots of things that can be done in different ways.

> There's this notion
that we need intelligence
services. Fuck that. <

Some people approached me saying they were with the Canadian Security Establishment. They were basically the NSA or the BND of Canada. They offered me a job, and I said that I wasn't interested.

They said, "Come on, you can have a passport."

I said, "No, I don't want to wait in line."

They said, "Don't worry, there is no line for you. Just come with us."

I said, "I really don't want to because you have a monarchy, and I'm disgusted by the idea of royalty."

They said, "The king has no power."

I said, "I don't want to prop up your monarchy, regardless. I'm an American and being born in America, I'm disgusted by hereditary power. I like the idea of being able to vote for a leader. I like to have a representative democracy."

They said, "How do you think Stephen Harper[32] makes decisions?"

I said, "I think you're going to tell me that you put some choices on his table and then he picks whatever you would be okay with."

They said, "That's right, we put three choices down in front of him, and he gets to choose between any three. There is no wrong choice. That is democracy, kid."

I said, "I want to work to destroy your world, because you are against everything that I have been raised to believe about free societies. You are subverting the fundamental trust that has been given to you."

32 *Stephen Joseph Harper is a Canadian politician and member of Parliament who served as the 22nd Prime Minister of Canada, from February 6, 2006 to November 4, 2015.*

They said, "The NSA signs off on every single one of our hires, so don't worry. No one will be upset with you in America if you join the Canadian version of the NSA."

I said, "Gosh, you can't say no to this question, because I keep saying 'no' and you won't let me."

They said, "Sleep on it, think about it, don't you want to change the world? This is your chance to change the world, kid."

The person I was with said, "Yeah, let's do it. Let's join up."

I looked at him and said, "I want to tell you about imperialism when we're not near these people. I want to talk to you about how fucked up this is."

We left and we talked about imperialism. We talked about how his country of Lebanon came to exist. We talked about how non-democratic actions drastically affect the world. In fact, how the way the Middle East is carved up right now comes from these non-democratic actions. That, to me, suggests that the democracy in Canada, according to these spies, is a charade. It suggests that some people know it is a charade and they are proud of it and use it as a recruiting tool.

Your mileage may vary here in Germany. My guess is that, since you haven't cleaned house with the Verfassungsschutz and after the NSU scandal, you probably have a similar problem.

Sounds like a structural problem. The NSA is divided into departments and subdivisions. There's a subdivision of customer relations, and you realize the customers are the White House and the FBI. Do these sub-organizations create their own structures, their own logic?

JACOB: I don't view myself as having a war with the United States. I view myself as using journalism to tell the truth about what is actually happening. In some cases, people wish the truth to not come out. They didn't want people to know that the NSA was spying on Merkel, for example. That, to them was important to hide. Injustices like the drone murders act as functions of the structure.

When you have the keys to the kingdom, you start to use them. When you have the ability to kill people without having a jury, without having a judge involved, you use it. The idea of the death penalty is so barbaric, but in the U.S., we have the death penalty in a lot of places. Naturally they go two steps further from that idea and the injustice becomes an emerging phenomenon of the structure and the organization itself.

By having an intelligence agency, you are sure that you'll have people who behave as if they're in a John le Carré novel[33]. People will always think they're above the law, and in fact, they are. They are living the modern John le Carré novel, and they're proud of it. That is a social cost. I'm hoping we can change the social cost.

Every time you find out someone works at the BND, ladies, and gentlemen, don't sleep with them! Raise the social cost for being a criminal spy so high that they quit their jobs and go do something more respectful.

Jacob, you were studying the information on surveillance by Edward Snowden. Could you describe your perspective on what you found most shocking regarding methods and techniques of surveillance on a hardware and a software level?

JACOB: In our world, there's a theory that your communications with someone else are spied upon. But actually, in our world, we have sensors, we have credit cards, we have cars. We have all these electronic systems that

33 David John Moore Cornwell (born 19 October 1931) (pen name John le Carré) is a British author. During the 1950s and the 1960s, he worked for the Security Service and the Secret Intelligence Service, and began writing novels under a pen name.

bleed data onto the Internet. There are also intentional communications.

The NSA, the GCHQ, the BND, the CSE, you name it—they try to collect absolutely everything, put it into databases and then remove the agent, in some cases, from the loop. They have algorithms that decide if you're a terrorist.

This kind of stuff is the scariest to me. Look at movies like *The Lives of Others*[34], with a quaint protagonist. He walks around and writes notes or sometimes doesn't write notes, because he wants to cover for them. He realizes that one note in one direction might be too much, and it's not actually what it would seem to be. That guy is completely removed from a lot of these systems.

When Jeremy Scahill and Glenn Greenwald wrote about the drone murder in *The Intercept*, they said they're actually targeting phones. They're targeting the metadata. If you happen to take my phone that day, and I was targeted, you get the drone strike, not me. Only afterwards, they realize they've killed an innocent person.

This is the reason that we have due process in the West, at least in theory. It's the reason we're supposed to have a trial by jury. What I found the most shocking is, because it is technologically possible for the NSA to help the CIA and the BND with these things, they just do it. Because it is done in secret, and there is no adversarial process, they just do it thousands of times and get away with it.

>This is a Philip K. Dick nightmare. <

34 *The Lives of Others is a 2006 German drama film about the monitoring of East Berlin residents by agents of the Stasi, the GDR's secret police.*

We all contribute to this planetary surveillance, and you can't really opt out. You can choose not to make a phone call, but you can't choose not to pay rent. You're always leaving some data behind. I don't have a cell phone, but it doesn't matter, because every person around me has a cell phone. This is the Philip K. Dick[35] nightmare world that we're living in. In theory, you have to worry about everyone being a spy, but in the Philip K. Dick world, you have to worry about everything being a spy.

Like the Internet of Things.

JACOB: Yeah, for example, like the Internet of Things.

You feel we're in the age of post-privacy? Is there a way back?

JACOB: We're not in the age of post-privacy. That is a rich-white-people paradigm that I don't understand at all. Ask a woman in this room if she has a double consciousness. How do I look? How do I look to an attacker on the street? We don't live in a post-privacy world because we're not in a post-privileged world. We are living in something that could be considered post-democratic, in some ways, or maybe post-civil-liberties.

In the United States, the idea of the Fourth Amendment is that unreasonable search and seizure is not allowed. But as Snowden recently pointed out, the U.S. government, and the German government, and many other governments around the world have decided that, in fact, seizure is fine. They can do it as much as they want.

When it comes to electronic surveillance in the modern world, this fundamental law is just gone. That's the case

35 Philip Kindred Dick was an American writer, whose published works mainly belong to the genre of science fiction. Dick explored philosophical, sociological and political themes in novels with plots dominated by monopolistic corporations, authoritarian governments, and altered states of consciousness.

for all of us in this room. Every person that has a cell phone, for example. You all have civil liberties in this room. The NSA, the BND, and the rest of these groups would say that those electronic devices you're carrying don't have those liberties. But that's indistinguishable from you.

Unfortunately, I think the whole post-privacy movement that exists is deluded. It's like a coping mechanism. It's like saying you're not allowed to wear these clothes anymore. We're just post-clothing, you know. Just deal with it. Just learn to love it. Learn to love not being free.

Fuck those people; they're crazy. I'm not me when I don't have a private sphere. I'm not me when I have to worry about everyone else constantly examining everything I do. That's post-existence, in some ways.

What do you say to people when they say, "I don't have anything to hide"?

JACOB: The whole notion of having nothing to hide shows shallowness in the analysis. For example, it's not about whether or not I have something to hide. I do have something to hide. Everyone has something to hide, but I also have a choice. Part of my ability to choose, part of my dignity, my agency is stolen from me when you know everything, when you predict everything, when you record everything, when you watch everything, when you rifle through all my papers, when you know where I look on the page of a book.

> To suggest that we're just talking about "having something to hide" is very naive, and also a matter of class privilege. <

Obviously, we all have something to hide. That is without question. The question is if we have a right to say, "No."

We should clearly say, "Yes, we have a right to say 'no.'" We have a right to privacy in our homes. We have a right to choose if we should be stripped of our clothes. We have a right to dignity.

To suggest that we're just talking about having something to hide is very naive, and also a matter of class privilege, in a lot of cases. That's not falling on the right side of reality.

Some people trade their private data for the convenience of certain apps. They don't give away the rights, but they do give away this information. Then, you have the big Internet players, who grab that information. How do you feel about these people, and what do you want to tell them and the producers of such apps?

JACOB: I refer to Facebook as *Stasibook*, because I think you're constantly reporting on your friends there. In many cases, Facebook has special relationships with law enforcement, and they give up that data. They wouldn't give it to me, but they will give it to the police, even illegitimate police. I think that's wrong.

We should build alternatives, but I also think we shouldn't fault people for living in the modern times. What choices do we really have? It is a privilege that I can exist without a cell phone, and not everyone has that privilege. We need to build alternative structures, we need to build alternative communication systems, and we should recognize, I think, when people have things like Facebook, it does not mean they do not care about their privacy. It means that the public sphere, a meeting place, like the public square, is replaced by an American corporation. People participate because there isn't an alternative.

If you have a social graph that spreads across many cities, or if you have family members that used to use the phone but now want to send a photo, what alternative do they really have for fulfilling that?

The reason Facebook is successful is because it does meet a need people have. That does not mean people don't have particular values. It just means they haven't yet found a way to express those things in a way that respects those values. We should persevere and build a system that does respect our autonomy, which does respect our liberty, which does allow us to share our lives and connect with people.

Angela, how did you change your perspective on privacy, public, and your usage of technology, since you were in contact with Julian Assange?

ANGELA: It changed a lot. At a certain point, I was censoring myself. I felt I could be watched more actively than everyone else. I instinctively wouldn't communicate openly with my emails. It was just a feeling. With just the possibility that somebody can read it, you automatically censor yourself. At a certain point, I decided to use encryption but, to be honest, it's a total pain in the ass. Sorry for saying that.

Basically, I'm too lazy, because it's still very complicated. It would be great if someone would invent a solution to make it easy for people to use. When I really want to exchange information that should stay private, I use encryption, PGP with a Key, along with some other things.

Actually, we organised the first Crypto Party in Cologne last time we showed the *Assassinate Assange* play. We did the Crypto Party in a state theatre. It was totally crowded. Many old people came, the typical bourgeois theatre visitor. And I was very surprised. We didn't expect that. They all wanted to know how it works. It's not even that complicated or difficult. It's more inconvenient, I'd say. We were all overwhelmed that there was a need for it.

When you go deeper into the topic of digital surveillance and encryption, you sometimes have the feeling you want to get off the grid, which is quite challenging because of social and convenience reasons. Jacob, you are deep into encryption features and all the surrounding tools—what is easy to use is insecure and vice versa. What is your perspective?

JACOB: RedPhone and TextSecure by Moxie Marlinspike are free Android apps, soon on the iOS, and both secure and easy to use. There're things like off-the-record messaging and PGP. I think PGP is the least usable, but it's very important, anyway. Of course, people can advertise a thing as being private or secure, when actually it is neither of those things. It is a problem. It is confusing.

> If you want to engage
in risky behaviour, you have
to protect yourself. <

Imagine this analogy: I grew up in San Francisco in the Bay area, in the '80s and '90s, when everyone was dying from HIV. I spent a lot of time in the hair salon that my parents owned. Many gay men that worked in this hair salon died from AIDS. One of the lessons we learned is if you want to engage in risky behaviour, you have to protect yourself. A lot of people refused to learn this lesson, because they thought they were exempt from HIV. They didn't want to wear condoms, for example.

There is also a societal part that is important, including raising awareness and building alternatives. Right now when it comes to communications, we have an epidemic. It's a serious epidemic. If you happen to be located in Yemen and you happen to be a Muslim, then it's a terminal illness, potentially.

We have to have education about this topic; we have to raise our consciousness. We have to constantly repeat this in public if we know about it. There is a war on your privacy. To paraphrase Bill Hicks, "Every time you use encryption and anonymity software, you're winning it."

The state also has to respect your autonomy. Your phone is a tracking device that makes phone calls that the state can install the state's Trojan on, for example. They can take over the core of your life, and they assert they have the right to do that. What is the difference between that and a criminal organization that takes over your phone?

I think you're touching a very important point. The government should be respecting our privacy. In Germany, we have something called Postgeheimnis (postal secrecy), and I wouldn't expect them to open my letters or my packages. I shouldn't have to use super glue for my envelope, giving the antidote to the receiver. So why do I have to do that in the virtual world?

JACOB: The state decided that they were going to violate your fundamental liberties and didn't consult you about it. We have postal secrecy in the United States, as well, but we recently learned that the Postal Service records the outsides of all letters.

Furthermore, in many cases, U.S. Postal Service workers can open your mail without a warrant. This has been revealed in the New York Times recently, because the postal service accidentally left a slip in a letter. The person who was targeted found the slip is his own mail, which basically alerted him to this fact.

Then, I revealed in *Der Spiegel* in December that the NSA is also doing this. When you buy a computer, they'll actually take your mail, open your mail, replace part of the computer, and then send it on to you. Why is this happening? It's happening because without transparency, government agencies like corporations have this sickness of power. They expand, and then they abuse it. It's really only transparency that will help us here.

Is rising technical literacy a feasible solution?

JACOB: It's one part of the solution. Like I said, if the state requires your phone to be wiretap-able, there is a trade off. Right now the trade off is that the BND and the Verfassungsschutz can spy on me. When you travel abroad, everyone gets to spy on you, and I get to spy on you here, if I want. We need a technology to match our social values. If it doesn't match the social values, and there's a discrepancy, you will lose.

>We need a technology to match our social values.<

Civil rights movements were always coming out of a niche and out of subversive momentum. At this point, it feels like the digital rights movement hasn't taken off yet, because it's not yet painful enough when your digital rights are violated.

JACOB: We shouldn't talk about Internet freedom. We should just talk about freedom. We shouldn't talk about digital liberty. It is just liberty. It's not about a digital right to protection against unreasonable search and seizure. It just the same liberty we have always had. It must not be eradicated merely because we use electronics. This does not mean that the fundamental underpinnings of liberal democracy should be destroyed.

Angela, you choose different stories. You choose a different approach. Is there a utopian vision that you can create, knowing that all these negative, repressive, and limiting things are out there?

ANGELA: I wouldn't go so far and say I can draw a utopia, because I have no idea of it. I would call it a kind of a movement toward enlightenment. As much as I can help *Aufklärung*[36] in this way, I'm trying.

36   The Enlightenment (also known as the Age of Enlightenment; and in German: Aufklärung) was a philosophical movement, which dominated the world of ideas in Europe in the 18th century. The Enlightenment included a range of ideas centred on reason as the primary source of authority and legitimacy, and came to advance ideals such as liberty, progress, tolerance, fraternity, constitutional government, and separation of church and state.

A Changing Narrative: From Hacking to Storytelling

And I don't think that people are not approachable. The opposite is the case. The play was always sold out from the beginning. Maybe for different reasons, like the scandal; but later, when Snowden happened, the status and image of Julian Assange changed. When I spoke with people in the beginning about him, they said, "My God, this is madman. He is totally paranoid." At one point, I was even thinking, "My God, maybe he's exaggerating. Total surveillance, really?" Now, it is proven through Snowden. People see Assange with different eyes.

> Unfortunately, we weren't pessimistic or paranoid enough. <

JACOB: We wrote a book called *Cypherpunks*. It was Julian Assange, Andy Müller-Maguhn, Jérémie Zimmermann, and myself. We talked about all of these issues, about freedom of movement, about freedom of economics, about total surveillance censorship, and so on. We examined these issues in a two-part video series, which was called *The World Tomorrow*. It was episode 70 or 89 or something. It's not highbrow intellectualism. It's just four guys sitting around talking. And we talked about these things long before Snowden, because people didn't want to believe it. Unfortunately, we weren't even pessimistic enough or paranoid enough. We were, in fact, quite optimistic.

People have cell phones. They run businesses based on the Internet. For people who are not so technically versatile, what should they do as a first step?

JACOB: First of all, I would encourage you to visit the piratebay.org and download a copy of our book *Cypherpunks*. You can also buy the book, but, hey, it's the Internet, so just download it. I don't make any money from the book anyway, and even if I did, I'd still encourage you to download it. It's called *Cypherpunks*.

It will introduce you to some technologies, like Tor, which is an anonymity network, off the record messaging. It allows you to encrypt your instant messaging and email with graphic PGP. This is the first step.

But further, you should recognize that when someone gives you the choice of subverting your communication systems and uses fear politics, like the fear of terrorism, the fear of money laundering, the war on drugs or child pornography, you should say, "No, I don't want to sabotage my systems. I want to have secure communications."

It starts with you. Examine your own world. Think about your Internet business, for example. Is it possible for your customers to communicate with you using the same level of privacy as if they walked into a shop? To do that, you really need to examine how this technology does impact your life.

You can also install apps on your cell phone, but security is really a process. Privacy is a process. Liberty is a process. You can't just install an app and be done with it. Nonetheless, you can install stuff like Techsecure or RedPhone or Private GSM. This way, you can start to have some privacy, integrity, and confidentiality in your communications.

Are you both still optimistic about the use of technology? Can technology help us or are we just fucked with technology, based on the whole process you saw?

ANGELA: I would say I'm a pessimistic optimist.

JACOB: I have friends who have died in the struggle and who are in prison. My friend Aaron Swartz[37] killed

37   Aaron Hillel Swartz was an American computer programmer, entrepreneur, writer, political organizer, and Internet hacktivist. He committed suicide while under federal indictment for data-theft, a prosecution that was characterized by his family as being "the product of a criminal-justice system rife with intimidation and prosecutorial overreach."

himself because of an overzealous prosecution, where they basically attacked him for downloading files that he had the right to download. They wanted to do that to him because of his politics. People like Julian Assange are much the same. I don't have a lot of hope for Aaron Swartz because he's dead forever. That is the reality.

>I have had friends who have died in the struggle.<

When I was at my friend's New Year's party, we had a toast, and I said, "We made it to 2014." A good friend of mine, Roger, said, "Well, actually, not all of us did."

I think that Laura Poitras, Glenn Greenwald, and I, maybe, will get out of this situation alive, but we will see. I don't have a lot of hope for myself, but I do have a lot of hope for each of you making a choice that tells us where our societies will go in the future.

I think that you'll make the choice regardless of whether or not you're conscious of it, and human society will continue on. It will be fine, I'm sure. I don't expect that some of my friends and I will be free. Julian is an example.

Look at most of the original WikiLeaks people. They either walked away or have been put in serious harm or had to break in a very serious, public way to gain back some safety and legitimacy, as far as the state is concerned.

I think I'm an optimist, despite all the things that I said that sound very sad. We still have some ability to change these things. There are still democratic structures that work. Germany's democracy is so much better than the U.S., in how you can meet a member of parliament and talk to them about your issues. I tried for four years to reach out to my congressperson, and they wouldn't have a meeting with me. It's not equal in how deteriorated it has become.

But Julian will probably die in the Ecuador Embassy. Snowden will probably die in exile in Russia. People in my country have actually said they should be assassinated. We live in a world where elected politicians openly talk about murdering their political enemies. They say it as a matter of pride and nationalism, and they are not laughed out of their office or arrested for inciting violence.

> To quote Edward Snowden: "The worst thing that can happen is that nothing is happening." <

2
What is the Colour of Money? The Myth of Economics

Tomáš Sedláček
Manouchehr Shamsrizi

PANEL DISCUSSION, BERLIN,
NOVEMBER 15, 2014

Economic models influence our understanding of reality. From the first shells
and wooden coins to today's crypto
currencies, we've always built our trust
in abstract values, which then define all
social interactions of trade and services
and determine the wealth of individuals,
corporations and even nations. In
Sedláček's contemplations, the invisible
hand of the market is as real as the
belief in god – and the extreme inter-
pretations of both are just as dangerous
to the global community.

Why is gold associated with security,
while Bitcoins are viewed as enablers
for crime and chaos? Which stories and
myths have led to our financial systems
today, and what opportunities lie in the
digitization of currencies and payments?

With the global economic crises, we look
to establish a better understanding of
the mechanics of economy and flow of

wealth. How can we trust in the powers that be, which have led to the unjust distribution of resources and which leave large parts of the world in despair?

How can new concepts of "sharing economy," supported through digital platforms and crypto currencies, change the balance of good and evil in economics?

PANELLISTS

Discussion leaders Tomáš Sedláček and Manouchehr Shamsrizi met one another at the St. Gallen Symposium, an annual conference that works to foster intercultural dialogues. The D.DAY panel allows them to discuss the history and future of money together, aligning their interests and intersecting their vast knowledge.

TOMÁŠ SEDLÁČEK became know through his work of economics, philosophy, mythology, theology and analysis of popular culture in his debut book *Economics of Good and Evil – The Quest for Economic Meaning from Gilgamesh to Wall Street*. The book has been translated into 16 languages and received the prestigious Deutsche Wirtschaftsbuchpreis at Frankfurt Bookfair. Prior to the publication, during his fellowship at Yale University, Yale Economic Review named him among the 5 Hot Shots in Economics.

A popular thinker, connecting a wide array of topics, he is a member of the World Economic Forum Global Council on Values, a member of the Advisory Board for Sustainability of KBC Banking Group, Chairman of the Supervisory Board of Vize 97, a charity organization founded by Dagmar and Vaclav Havel, a long-term member of Czech National Economic Council (an advisory group to the Prime minister) and many others.

MANOUCHEHR SHAMSRIZI, who was chosen as a "European Future Leader" and a "Leader of Tomorrow" (St. Gallen Symposium), is widely considered the voice of a new generation in Germany. He has spoken at the German President's Bellevue Forum, the Malteser Future

Laboratory, re:publica, TEDx and Swiss Re's 150-year anniversary.

In parallel to his studies, the North German of Iranian background became the youngest Global Justice Fellow at Yale University, an honorary member of Peterhouse MCR at the University of Cambridge and was invited to join international think tanks like Google's CoLab, Siemens' FutureInfluencer, ZEIT foundations beta-group, Howard W. Buffets Global Impact Institute, Wilton Parks Atlantic Youth Forum and Muhammad Yunus' Grameen Creative Lab. Manouchehr co-founded Kryptos, an NGO focusing on the benefits of crypto currencies, and has been a vivid speaker on this topic at international conferences over the last year.

>All of our values have become a subset of GDP.<

Tomáš, you wrote the book *Economics of Good and Evil*, which I really enjoyed reading. What leads to an economy of good and evil—why this duality?

TOMÀŠ: I think in the age past we have had the tendency to view economics as something that is divine. It was perfect, it was self-regulating, and it was omniscient, omnipresent. It had the keys to prolific rationality.

It became what I call "the unorchestrated orchestrator." You cannot orchestrate it—laissez-faire, let it be, don't meddle. It has its own logic. It has its own rules that you cannot even comprehend. Only the priests can decipher the signals that it sends. By priests, I mean economists. They are the only ones who can interpret the whims of these deities. So you cannot orchestrate it. It must be unorchestrated. It will orchestrate you.

It will tell you what the meaning of life is. It will give you values. You can see that even today. All our values have become a subset of GDP[38]. But what I hate the most is the debate on the creative industry. I even hate the connection of creative and industry. That is an oxymoron. Nevertheless, the only rule, the only meaning of art—this is the debate that is going on in Europe, and I imagine this might be quite alive here in Berlin—is, that art is legitimate, because it adds to GDP. That's the only way that we can legitimize art.

But in Oscar Wilde's famous foreword to *The Picture of Dorian Gray*, he ends with the sentence: "All art must be quite useless." By useless he doesn't mean meaningless, but that art is exempt from the imperative, which is

38   *GDP = Gross domestic product*

otherwise predominant in our lives. That is, the imperative of "you have to be useful, you have to be efficient, you have to be contributing."

Has this imperative aspect an evil dimension?

TOMÀŠ: No, since art is exempt from that. A little bit like Nietzsche said, this God of economics is dead. This God is non-existent; there is no invisible hand. There are only our hands, right hand and left hand, and they can do good and evil.

Economics, like any other institution, be it democracy, be it the rule of law, be it journalism, be it anything, can be both. It's not neutral. It can be good and evil. You cannot rely your moral judgement on a system, which we call capitalism. If I could say it better, I would not have written a book of four hundred pages.

Manouchehr, as a student of Dirk Baecker[39], you also like to think things through from a meta perspective and try to have a holistic view. I would love to hear your understanding of good and evil or your inner compass?

MANOUCHEHR: I'm not only a student of Dirk Baecker but also of Tomáš. And Tomáš just pointed out the invisible hand, a term used by Adam Smith[40]. There is another famous concept of Adam Smith, which is usually wrongly translated into German. It is the term invisible judges, which is often translated into "die unsichtbaren Richter." But in Adam Smith's original paper, this term was described as an inner moral compass. And Smith was pointing out that this is something that everybody has to come up with by him or herself.

39 Dirk Baecker is a German sociologist and holder of the Department of Cultural Theory and Management at the University of Witten/Herdecke.

40 Adam Smith was a Scottish economist, philosopher and author. He was a pioneer of political economy and a key figure during the Scottish enlightenment.

Taking that idea further, my moral compass is quite flexible. That doesn't mean that it is constantly changing by 180 degrees. But I'm trying to accept complexity, by which I mean that I am conscious about the fact that I can't be sure about the final outcomes of my momentary actions. I'm also not sure if it will meet my moral standpoints. I cannot be sure if the path I chose will lead me there.

In preparation for this talk, somebody asked me; "You are having Tomáš Sedláček, so why are you talking about crypto currency? What is so digital about Tomáš?" I said, "We will talk about the history and about the present of money. And if we talk about the future, this is something we have to take into consideration."

Talking about the future: Manouchehr, you were involved with crypto currencies through your NGO Kryptos[41]. How did that happen and what is Kryptos about?

MANOUCHEHR: Kryptos is Germany's first NGO looking into crypto currencies and the technology behind it, and the impact that the technology can have on philanthropy and society. We are not focusing on a specific currency. We are not earning money by anything we do in that setting. It is a spin-off of The Global Shapers[42], the youth organization of The World Economic Forum[43].

Why am I interested in crypto currencies? Many reasons. One goes back to the novel *Lord of The Rings*, which was an idea I got from Tomáš. If I want to understand what economics will look like in 100 years, I should understand how the Elves in Tolkien's *Lord of The Rings* live. They have very little emphasis on material things, because all things they produce materially are of very

41   Kryptos is Germany's first NGO, which analyses crypto currencies and their relevant technology

42   Global Shapers is a network of city-based Hubs developed and led by young leaders who want to serve society.

43   The World Economic Forum engages the foremost political, business and other leaders of society to shape global agendas.

high quality. They don't need to renew them all the time, which gives them time, space, creativity and resources to look for visions, ideas, and new economical settings, which are built on different ideas and dreams. Crypto currencies as a digital version of money are logical steps into this highly efficient elfish economy.

TOMÀŠ: It's a great pleasure to become spokesperson of the Elves.

MANOUCHEHR: I agree, much better than for the Ogres.

Economist Hyman Minsky[44] once said, "Creating money is easy. The hard part is getting it accepted." Tomáš, how do you see the role of money changing?

TOMÀŠ: In my view, every currency is a crypto currency. The Euros you have or the Deutsche Mark you used to have is really a proxy. If I can go very quickly through the history of money in two minutes and I'm going to try and make it interesting.

> Every currency
is a crypto currency. <

In the beginning, we know that the oldest money equivalent was actually written on clay. It wasn't gold or any precious metal. Today's money is plastic. Most of your money is somewhere on a plastic card. And money isn't me. It only exists in a relationship. If I establish my own currency, I could easily become a millionaire or trillionaire, but that doesn't really count. Money is basically fundamentally a token of relationships.

If you want to confuse an economist ask him or her what money is. Nobody really knows what money is. There was

44 *Hyman Minsky was an American economist and professor of economics at Washington University in St. Louis.*

the German philosopher Georg Simmel[45], who wrote a very long book about money. After reading that book, you still will not know what money is. Most of your money actually isn't even in paper. This is very unoriginal lithography. It is in a bank. It is a digital record of ones and zeros.

And you may have 200 euros maximum on you, but the rest of your money is somewhere in the bank. And if I want to send you 10,000 euros, it gets transferred from my account to yours, assuming that I have 10,000 euros in my account.

And this is the nice thing about economists: we can assume anything. It is a little bit like *Star Wars*[46]. When I was younger, I used to watch *Star Wars* with my father. My father is a technical type of guy. He is a pilot. He would watch it and say, "That is not possible. That would never fly. It is a laser blaster that means the laser blasts will travel at the speed of light. Which means you can't see them and you can definitely not dock them." He had many very good points.

So I realized that we call *Star Wars* science fiction, but Economy is social-science-fiction. If you want to watch *Star Wars* or *Lord of The Rings*, you sort of have to accept that Elves exist, that travelling with Chewbacca is possible. You really have to do a belief exercise and forget about all the rules of physics. Otherwise, you will not be able to watch it for more than five minutes.

> Currency is a proxy.
It is an agreement that we
have between each other. <

45   *Georg Simmel was a German sociologist, philosopher, and critic. He was one of the first generations of German sociologists, asking, "What is a society?" in direct allusion to Kant's "What is nature."*

46   *Star Wars is an American epic space opera franchise centred on a film series created by George Lucas.*

And a very similar psychological procedure is valid, when you enter the realm of economics. There are also many things you have to believe. For example, you have to believe, that human beings are rational. You have to believe in money itself. This is a belief of our time, which is sort of channelled into a symbol, like Jesus Christ, when God becomes a material human being. This is abstract. Trust is becoming physical. At the end of the day, it is all a crypto currency. It's a proxy. It is an approximation. It is an agreement that we have between each other.

Let's talk about trust in relation to crypto currencies. Through the last financial crisis in 2008, we experienced a big loss of trust in today's money systems, especially with the bank bailouts. With this event, crypto currencies were on the rise. How do you see the relationship between money and trust?

MANOUCHEHR: First, I see trust as a built-in notion of social communities. And different social communities have different things they trust. For example, if you analyse *Star Wars* and *Star Trek*, you will see that there is a strong political belief system represented in each of the stories.

Star Wars is about a group of rebels who fight against a tyrannical evil empire. The two groups fight each other, but paradoxically, both the rebels and the empire are based on the same imperialistic ideals from the past. Compared to this, *Star Trek* is made of stories of mankind from a future where national, ideological and ethnic divisions are a thing of the past, and where learning and humanism are upheld as the highest values. We can see that the story of *Star Trek* is mainly addressing a more liberal life view, while *Star Wars* is very much about power structures and therefore promotes a more conservative ideology.

If you follow Dirk Baecker, trust is the only thing you can trade. Banks do nothing else than offer you trust, and you pay for it. But if you don't trust the intermediary in the system, why should you pay him? The idea of the trusted intermediary is very old. But today, we have technologies to rethink a centralized bank system.

Back in the days on little islands in Mikronesia, they used Rai Stones[47] as a payment method. Every family got one stone and then marked on each stone what you traded with your neighbours. This system was already working without an intermediary.

This is actually the basic philosophy behind crypto currencies or Blockchain. You're meant to exchange value directly, without using a trusted intermediary. You build up a person-to-person trust through a crypto currency system that allows you to track transparently who owes what to whom.

Many economists once stated that humans started with a commodity-based money system. David Graeber[48] says that this is not true, and that we are living in a political-based currency system today. You don't trust the physical money, but rather the political system, which forces me to pay you back what I owe you. And with crypto currencies, we now might even go one step further and enter a math- and algorithm-based way of trusting each other. It is trust in a belief system; in a new religion of algorithm, which can be seen very critically as well.

But in a village, you know each other, and on the Blockchain, you are kind of anonymous. Everyone is just a number. What does that mean as a consequence?

MANOUCHEHR: Yes, and in the crypto world, I do not trust you in person; but I trust the numbers.

47 Rai stones, or stone money, are large, circular disks made from limestone, used as money on various islands, including Guam.

48 David Graeber is an American-born, London-based anthropologist and anarchist activist, perhaps best known for his book, Debt: The First 5000 Years.

Which means you rely on total system control.

MANOUCHEHR: Yes, and I trust a system, which is open source and decentralized. I trust the fact that my computer is constantly checking on algorithms, along with millions of other computers in the network. There are around two million people using different kinds of crypto currencies today. And if one of those people creates a new currency and a transaction with this new money should be processed, it can be rejected by all the other computers in the system. This would happen if they came to a different mathematical conclusion on the basis of some algorithms.

Tomáš, what do you think about trust in relation to money?

TOMÀŠ: The question is: is there a boycott of trust?

I had this very similar debate in 2009, with this great thinker Oliver Tanzer[49] in Krams. The host asked us: "Is there nothing to trust in anymore?" I said, "It's exactly the other way around. There is everything to trust."

The problem is exactly as you were saying: we now know there is everything to trust and everything to believe.

>Money is, in a way, automated trust.<

Money, for example, only works if you believe in it. The problem in my reading of the world isn't that there was too little trust. No! The problem was that there was too much trust before 2007 and 2008: a kind of blind trust.

And all trust, in the end, is blind, and so is money. If trust is blind, then money is perhaps also blind, following that

49   Oliver Tanzer is an Austrian journalist and senior editor for foreign policy and economics at Furche.

logic? Our problem was a kind of religious one, and I have nothing against religions. I believe in everything. I even believe in UFOs. I believe in *Star Wars* and in *Star Trek* at the same time. But the real problem was that we trusted too much. Money is, in a way, automated trust. You don't have to examine whether I trust you or not. Money automates that for me.

When you buy a painting by Damian Hirst, for example, the value isn't in the painting; the value is in the piece of paper, which says, "This was painted by Damian Hirst." It's a trusted source, straight from some trusted expert on art or Damian himself. The value of the painting, the aesthetic value of the painting isn't really the value component of it.

That's why we have all these holy symbols on money, all except the Euro, which I love, because for Europe, it is almost impossible to find common holy symbols. Most of the holy symbols are incomprehensible for the other Europeans. That's why you find bridges and more neutral signs on the Euro.

Normally, a nation would put its most holy symbols on its currency, like capitals, historical figures, saints, or the signature of the chairman of the bank. That is the only way to make that piece of paper valuable, with all holy symbols we possibly have. We put them all over our bank notes to make them trustworthy.

Besides value and trust, money and currency are also enablers. Right before the start of the panel, we discussed that the Islamic State (ISIS)[50] is trying to issue its own crypto currencies. What would that mean? On one hand, you have crypto currencies evolving, and then you hear about the Islamic State issuing its own. How do you see that?

50   *The Islamic State of Iraq and the Levant (ISIL, also known as ISIS), is a Salafi jihadist militant group that follows a fundamentalist doctrine of Sunni Islam.*

TOMÀŠ: I love the idea about money being a political contract. Meaning, who would use the money of ISIS, then?

The beautiful thing is, most of the money things are written in the very language, itself. For example, we call the crisis a credit crunch. *Credit* in Latin means trust, coming from *credo*. Those of you who go to masses know that. You don't even have to translate it: it is a *faith crunch* and that's what we've been calling it all these years. What we believed before is what we can no longer believe. It is a sort of a protestant revolution, without Calvin[51]. Meaning, we no longer want to accept that the Catholic way of approaching God is the only way. We know this, but we don't have any figure to follow. We are looking for a new Calvin or a new Luther.

> You are free to come up
with your interpretation of
money based on an algorithm
you choose. <

MANOUCHEHR: Or we are looking for many Calvins and many Luthers, all at the same time. Because if money is blind, as you just said, the real question is, whom do you choose to interpret faith?

TOMÀŠ: Or whom do you allow you to tell you your story? The most important thing in our lives is: who will be telling you the story?

MANOUCHEHR: The beautiful thing about crypto currencies is that we are now free to listen to as many different stories as we want. It enables everybody, including ISIS, unfortunately, to come up with a story. And we have seen this with different types of crypto currencies.

51   *John Calvin was an influential French theologian, pastor, and reformer during the Protestant Reformation. He was a principal figure in the development of the system of Christian theology later called Calvinism.*

Crypto currency as a technology and as a vision is much broader than just Bitcoin. Bitcoin has a very libertarian way of interpreting and telling the economical story, while, for example, Freicoin[52] is following a specific school of thoughts on coins. They came up with a coin, which automatically loses value over time, because they believe that this will build a more stable economy.

You are free to come up with your interpretation of money that tells a certain story based on an algorithm you choose and want to be spread. You might end up having parallel storytellers. And on a crypto currency marketplace, you can freely choose the story you prefer.

You are telling the story that crypto currencies are open, open source, maybe even egalitarian, in a way. But when you consider players like Warren Buffet, who once said, "The secret of getting rich is to be greedy when others are fearful and to be fearful when others are greedy." Is the story all about greed and fear? How much space for that story is in the Blockchain or for individuals to take advantage of?

MANOUCHEHR: There might be little space in one but a lot more in another Blockchain. Think of it like MySpace[53] versus Facebook[54]. What we are discussing right now regarding Blockchain is what Bill Gates[55] said in the early '90s about the Internet. We had no idea what the impact of the Internet would be, but we knew that the impact would be huge.

52   Freicoin is a peer-to-peer digital currency delivering freedom from usury. It's based on the accounting concept of a proof-at-work block chain used by Satoshi Nakamoto in the creation of Bitcoin.

53   MySpace is a social networking website offering an interactive, user-submitted network of friends and personal profiles. It was the largest social media network from 2005 to 2008, when it was overtaken by Facebook.

54   Facebook is an American for-profit corporation and online social media network. It was launched in 2004 by Mark Zuckerberg. It is the most popular social networking site in the world.

55   Bill Gates is the co-founder of Microsoft and an American business magnate, investor, author, and philanthropist.

> In every feeling of fear lies a little bit of desire. <

We don't know if Bitcoin will become the next MySpace or Facebook, but we are first interested in the philosophy and the mythological impact of the idea of a social network. You have to differentiate that. We don't know what kind of crypto currencies can be built, or what stories can be told with this new technology yet. We are right at the beginning.

Earlier, you said there is no longer an invisible hand. Maybe we agree that Adam Smith's invisible hand of the market is not as invisible as he thought. What you are describing, however, is a mathematical process that is not clear to everybody, only some. There must be invisible powers at hand. What do you think they are?

TOMÀŠ: There are always invisible powers at hand. For example, love, beauty, things that are not you.

But then the question is: why can't it be love and trust? Tomáš, once you said that money is more like energy. I think this is also related to that question. And the question is: what happens when this energy is triggered by fear and greed?

TOMÀŠ: I think the key to understand fear is to realize that in every feeling of fear lies a little bit of desire. In every desire there is also a little bit of fear. These emotions never come clean. You are falling in love. You want to be with her. You are indulged in desire, but, of course, there is this extremely erotic, beautiful component of fear and vice versa. You watch a horror movie because you want to enjoy fear, but there is also a strong component of desire for that. The dynamics of that have not yet been clearly described.

Again, let's go all the way back to the beginning, when we as mankind learned to write: there was *Plato*. He had this prophecy that spirit will emancipate itself over matter. This is, if I may say, a little detour, but actually completely related to the digital era. At this time, mankind is moving from this physical earth to the world of digital. You see it around you. In the prehistoric times, there was the great "Völkerwanderung" or "Migration Period"[56].

> We will move into
the realm of abstract
virtual reality or better:
real virtuality. <

This is happening again, but we as mankind are moving into the digital world. Just mark how many of us are already in the cloud. Funny enough, this is also something that Jesus spoke about, regarding heaven. We have this image that things are in the cloud. We are moving there. It's a fantasy world, just like money is a fantasy world, along with *Star Wars* and economics and everything. It's a fantasy world into which we are slowly moving. And a lot of people resist this. They don't like the idea. They say, "No, I want to touch things." I even think that in 10 years, shoe production will go bankrupt. We will move into the realm of abstract virtual reality or as philosophers call it, real virtuality, which I think is a nicer way to put it.

But what happens to the material world then?

TOMÀŠ: It will be like a forest. We come from the forest. We all know that. Today, you go to the forest once, twice a year, to sort of relax, but we don't live there anymore. Our habitat is in cities. And in maybe 10 or 20 years, in the same way, we will go to reality. It will be like going to the forest today, once, twice, three times a year, just to feel more real.

56 The Migration Period was a time of widespread migrations within or into Europe in the middle of the first millennium A.D.

Crypto currencies are maybe a leading example of how we are moving into the fantasy world, where everything is possible. You dream it. It will be real.

MANOUCHEHR: Another example, which shows how strong the correlation between mythological belief systems and modern technology is, is the whole concept of *longevity*, which Google and others are pushing with billions of dollars. This philosophy on longevity is in a very early stage and as part of transhumanism[57], it claims that the next step for humanity will be that humans live their daily lives in a virtual reality, in which you are immortal. This is also an old idea, as we learn from Tomáš's book, since actually Gilgamesh[58] is looking for immortality. So in 20 or 30 years you may just continue living as an avatar after you died. I can't say if this is good or evil, but it is something.

TOMÀŠ: How many of you in the audience have tried virtual glasses? You know how immersive and wonderful this world is, even if it is very lonely. But it is a world where you can travel in time. You can travel in space. If a comet hits us, we will just download you.

> Already now,
I could probably download
80% of you. <

Already now, I could probably download 80% of you. I know your desires ... or Google knows ... your preferences, your money status. I don't like numbers, but I guess

57   Transhumanism is an international and intellectual movement that aims to transform the human condition by developing and making widely available sophisticated technologies to greatly enhance human intellect and physiology.

58   Gilgamesh is the main character of the Epic of Gilgamesh, a poem that was considered the first great work of literature. He's a demigod of superhuman strength who builds the city walls of Uruk to defend his people.

60% of us are already in this world with one leg. With the other, we still sort of like to be in the real. Otherwise, if a planet hits the earth, no problem, we can just send a data bank.

In my most abstract imagination, we will leave a body behind and we will move, we will upload or download—like heaven or hell—into this digital world. This is also what old philosophers say. The question is, is it possible to take a human soul and move it to the digital world completely? You would live in your motherboard then. It's actually funny that we call it *motherboard*.

MANOUCHEHR: I am waiting for the Catholic Church's view on the theological aspects of transferring souls.

It's like reincarnation, but through a program.

MANOUCHEHR: Or teleporting. Can you teleport a soul? Can you upload a soul?

Tomáš, you write in your book, "We have given too much power to mathematicians and lawyers, taking it away from philosophers and writers." How does this perspective fit with a crypto world based on pure mathematics and algorithms?

TOMÁŠ: Payback time! For those of you who have seen the virtual world, it really is a world. To explain it to those who don't, the way you look into the virtual world is: you put on *glasses*. You can actually even do it with Google Cardboard[59]. You put your cell phone in the Google Cardboard, in front of your eyes, and turn on a certain virtual reality app. The Google Cardboard glasses are magnifying glasses that make you see the three-dimensional world. You look around you, and you can define where you are.

59 *Google Cardboard is a virtual reality platform developed by Google for use with a head mount for a smartphone.*

Art will go there immediately. It would be a wonderful debate, how art and technology follow each other. In the beginning, art was paintings with flowers and human beings or whatever, which are representations of the real. Today, nobody paints flowers. So art itself transforms more and more into the abstract. Now they're creating abstract triangles, and everybody says, "Impressive! That is a fucking triangle." So art is moving completely into the world of abstract.

There must be some meaning in the art piece, of course, but that meaning is never interpreted. Unlike the scientist, the artist must never interpret what he or she actually meant. And there are good reasons for that.

> In art, you will
soon be buying worlds,
instead of paintings. <

But in art, you will soon be buying worlds, instead of buying a painting. The difference between virtual reality and the world as we know it today is, that all our fantasies had a frame. A painting has a frame. A TV has a frame. An iPad has a frame. Everything has a frame. And with the frame it only represents 5% of your visual scene.

In virtual reality or real virtuality there are no frames. So when artists create a world for you of Elves or Goblins or some fly-through spaces, there will no longer be frames. It will completely escape that.

MANOUCHEHR: Is anybody familiar with the Brain Orchestra[60] project? That is a lovely project in which they use virtual reality aspects to enable people to compose

60   The Brain Orchestra Project shows the raw, creative abilities of the unmediated brain. It explores the question of what creative content the brain can generate that can ultimately bypass their bodies and become physical, out in the world.

and play music without the need of learning an instrument. You can argue that the quality is not comparable to the old masters, but still, you are empowering people to enjoy music directly by giving them the technical tools. Those instruments are the weirdest. They don't look even close to anything we know as an instrument. But why should we only use the former technology to express ourselves? It's like only being allowed to use money if we are qualified with Mastercard.

As much as I believe in it, your fantasy world offers a sort of elitist conversation, especially in looking at where the wealth is and how it's distributed. Tomàš, in your book, you say where money is going, and how it's accumulated. How do you think this accumulation can change in the next 20 years in a world with accessible digitalization? We have surpassed the body-centred economy into a mind-centred economy, where science and ratio dominate. Maybe the next thing would be an emotion-center economy?

TOMÀŠ: Yes, absolutely! We are entering an era of emotions. It used to be an era of material stuff, like axes and arrows. That's how we used to dominate each other. Then there were two or three hundred years of mind, with philosophers, and lawyers. A lawyer is actually a perfect example of a technical instrumentalization.

But what I wanted to say about artificial reality in this regard is this: You get digital images into your eye, and that digital image is a wormhole into yourself. That digital computer is only a way of structuring your imagination. I would say it's not structuring your thoughts, but structuring your imagination. That cell phone, TV or whatever you are watching isn't really a fantasy space. It is a structured civilizational gateway or wormhole in your pocket that helps you organize.

For instance, let's say you are in love with somebody. You close your eyes, you try to imagine him or her, but you cannot because your imagination is too limited.

Perhaps you remember that he or she has pointy ears and sharp teeth, but you cannot hold or maintain other details in your fantasy. This is a very strange component, which I think will become very important in the future debate about this abstraction.

In fact, we need tools to help us structure our imagination. For example, even if I am rolling, I cannot really tell the whole *Harry Potter*[61] story in my imagination. I imagine something, but I have to write it down. I have to have a pen. I have to have a paper—a lot of paper. It's the same way with a computer program. I am helping my imagination, structuring it in some ones and zeros.

Artificial reality is our imagination, as wonderful as it is, as it creates a dream world. I'm quite religious. I would like to see the *Apocalypse of St. John*[62] or the final book of the Bible in virtual reality. That's possible. I could travel the seven heavens and seven hells in virtual reality.

MANOUCHEHR: There are high schools that are using *Assassin's Creed*[63], a game that is built on brilliant graphics, by today's standards, for teaching. I mean, as a student, why shouldn't I be able to travel and learn about old Rome or whatever, without being there?

I think in 2004, or maybe even 2002, there was *Virtual Troy*, a project where, as I remember, some scientists worked to explore these possibilities for educational tools. Again, these are tools, meaning somebody is going to make money. How do

61   *Harry Potter is a series of fantasy novels written by British author J.K. Rowling. The books have found immense popularity and commercial success worldwide.*

62   *The Book of Revelation, often called the Revelation to John, the Apocalypse of John, or simply Revelation or Apocalypse, is a book of the New Testament that occupies a central place in Christian eschatology.*

63   *The Assassin's Creed is a franchise centred on an action-adventure video game series developed by Ubisoft. It depicts a centuries-old struggle between the Assassins, who fight for peace with free will, and the Templars, who desire peace through control.*

you think we'll get away from that collective emotional society towards a kind of global cohesiveness?

TOMÁŠ: This is something interesting that Slavoj Žižek[64], our favourite, absolutely crazy but genius thinker, says about Coca Cola. He says, "Coca Cola is a perfect communist drink." You can be rich as a pig, but you can't get better Coke. The workingman or workingwoman, the poorest of our society is drinking the same Coke like Warren Buffett. Unlike wine, where you can actually go and spend ridiculous amounts of money for an old bottle, there is no social stratification, if you will, in Coke.

>Coca Cola is a perfect communist drink. <

In the digital world, you can't really have a better iPhone than almost everybody can afford. Your gateway will be extremely cheap. Like if you buy two kilos of potatoes, you get a cell phone for free. In a way, if I take this imagination about real virtuality all the way, you can live in a mud hole, and you put on your glasses, and you can "transform" your building to be made of gold, diamonds, silver, etc. The traditional values in the real world will absolutely disappear. You can have a palace with 300 or 5,000 rooms. Why not? It's just a program.

MANOUCHEHR: You can beautifully combine the idea of the gateway and the trust topic. If you structure human history in the way system theory thinkers did, you will end up with the assumption that the technologies we use to communicate ultimately structure our organizations and institutions. As long as we only were able to speak to one another verbally, villages were the societies we built, because in the village you can still—through the gateway of speaking—share your imagination about things.

64　*Slavoj Žižek is a Slovenian psychoanalytic philosopher, cultural critic, and Hegelian Marxist. He's a senior researcher at the Institute for Sociology and Philosophy at the University of Ljubljana.*

Once we started to writing on paper, we could come up with the concept of a city, because we owned a way of using the new communication to build up more trust and new forms of governance. Once we had Gutenberg's printing press, we were able to print and had the opportunity to quickly spread information on different parts in a nation. That's why we came up with nation state.

Now, we are at the cutting edge of what we call the next society, the digital-based society. If you take the notion of the gateway further, and the evolution of technology in regards to human organizations, we see the ease for more and more people to share and access information globally today; almost in real-time.

Of course, there is still a digital divide[65]. But there is also a trend of democratization regarding the access to information. Until a few centuries ago, it was much more difficult to get a specific book, if one only existed in two copies: one in Kloster Bad Wimpfen and another in Hamburg. The gateways are providing easier access to information for more people.

You can also look at that from a psychological perspective—Jeremy Rifkin[66] is doing that a lot. You can discuss if Jeremy Rifkin is cool or not. But his idea about global empathy is pointing out that we first feel empathy with our fellows we can directly talk to, and then we feel empathy with our own nation state. Still, with digital media a global empathy can evolve today faster than ever before. Today, when Haiti is happening via Twitter, people start to spend money in a philanthropically cause within minutes. And crypto currencies even foster that, since the transaction costs are extremely low, which serves probably a global cohesiveness.

65 *A digital divide is an economic and social inequality with regard to access to, use of, or impact of information and communication technologies.*

66 *Jeremy Rifkin is an American economic and social theorist, writer, public speaker, political advisor, and activist.*

Coming back to your escapism, what do media do? It's similar to a description of the Holodeck[67], to some extent. Media has the role of escapism. So how do I get to be involved with society and where does this society exist? You create your elves society, I create an Orcs[68] society, and others create different societies. In that world, we all don't exist together anymore.

TOMÀŠ: That's why I said that's a lonely world. I can invite you for coffee or for beer, but I can't really invite you into my virtual reality, because we don't own it together. If I create my own world it will be a lonely world where I can define you, even if you're not there. It's a little bit like in *Matrix*, the scene in the second part with Trainman when he says to Neo: "You don't understand. This is my world. I built this place. I am God."

Before, I didn't understand that *Matrix* was a prophetic film about the future of mankind. The only thing is that we will enter that world voluntarily. People say they won't. But how many people don't have a smartphone? Almost everyone in this world.

This event room is also a world that somebody created for me. This room and this house was created by a brilliant architect. And we are in an absolutely unnatural position right now, because we are actually 30 meters above ground. We are actually "floating" in air. We have no business being up here. We were supposed to be 30 meters lower. This whole house, and the roads, and the microphone, and the cell phone has been manufactured, laboured and paid for by other people.

Those of you who were born in the city have never in your life seen anything natural. Look around you. All you see is artificial things, made by human beings for human

67   A holodeck is a fictional plot device from the television series Star Trek. It is presented as a staging environment in which participants may engage with different virtual reality environments.

68   Orcs are a race of creatures used as soldiers and henchmen in J.R.R. Tolkien's fantasy writings.

beings, artificially. Even the trees in the city are there because some bureaucrat thought it would be nice if that tree is there. It has nothing to do with nature.

MANOUCHEHR: This has lovely implications, because you can make a great analysis on how the city you were born in or moved to is structured and this way form your consciousness, concept and philosophy about nature. We think of nature as something that we want to keep. We want to have it. We want to protect it. In the last 200 years, it was known as a place where you go once or twice a year to relax, something beautiful.

But none of us has been attacked by a wolf in his lifetime. If you go to Japan, you see in their theological foundations, the ghosts of nature are aggressive and dangerous to humanity, and you have to stop them. That's again something thought of as being either good or evil. It's a bias.

When you say that economies and our reality are already very virtual and have been for decades, where are we in 100 years from now? Is there another level of virtuality?

TOMÀŠ: I don't know. Business and money are based on the idea of scarcity. If things are not scarce, if you copy, and paste, you can make not one room, but thousands of rooms or millions of rooms. Then what will we trade in the future? I don't know, as I said, but I think we'll be trading fantasies.

Artists will become the rich upper class in this world. If you've seen the *Lego Movie*, you understand. They will be the makers, the masters, or the crafters. Artists today are different. I like Manouchehr's example of being a musician without the trouble of actually learning how to control your fingers over the keyboard. Now you can be a musician without ever studying music.

>What will we trade in the future? I think we'll be trading fantasies.<

In the very near future, you can be an artist without having to know any history of art or some technique. Art will emancipate itself very quickly from techniques.

My guess is: we will be trading emotions. We will be trading fantasies. They are also quite closely related.

MANOUCHEHR: One of those emotions you would be able to trade would, of course, be the total antithesis of what we're saying. There is a great market of people offering classical piano in a totally offline setting. You'll be able to have that as well. But then it's a choice. It won't be the only way of being able to enjoy art.

TOMÀŠ: It's interesting how magnetic this world is. In theory, you can switch off your cell phones for a day or two or for a week, but you don't. You simply do not. This digital world has a gravity of its own.

I would even go further. I have never said this loud. I even think that the digital world has something, which make you sort of want to touch it. You have a touch phone; you want to touch it. You know there is nothing new happening on your phone, but you just open it and you touch it. It sort of wants your attention in the same way that a pet, a dog, a cat or a partner wants your attention. This digital world is not neutral. It is not without gravity of its own.

Just to finish my point on this: if you were born in the city, you've never seen anything real. There is no escape. You think, "I will buy a two-month visit to Tibet, and I will leave my cell phone here." But, no! You're still only doing a two-month visit to Tibet with your credit card loaded with German Euros at your disposal. You cannot leave this world of artificial virtuality.

One final point: This is the difference between nature and civilization. The whole idea of civilization is to detach us from nature. This is where Gilgamesh starts. He builds a wall around the city to separate from nature and to create this artificial world.

For those of you who enjoy horror movies, the leitmotif of horror movies is that nature reverses the food chain. Stupid horror, which I love very much, is *Texas Chainsaw Massacre*. The idea it plays with is this. We go like goblins[69] into the ground and extract stones that we torture under heavy temperatures. We create iron out of that. And out of that iron, we create a chainsaw, and we use that chainsaw to cut down living trees. Then we further torture the trees. We cut them into small little lengths and then dry them and turn them and twist them. Then you have a floor. Just in this room, we see extreme levels of violence, which is extremely silent. This room has been made by tons of violence performed against nature.

Read the very old texts. When our fathers were plowing the field, they felt extremely bad about it because of the violence. You take a piece of iron, you thrust it into the ground, and you tear it by plowing. This is why they have all these rituals for nature.

Often, nature in horror is represented through an animal, a zombie, a vampire, a werewolf, or a witch—who always lives in the forest—or even a small child. Today, a very typical representation is a madman, like in the movie *Seven*, or a person who is devoid of civilization or rationality.

69   A goblin is a monstrous creature from European folklore, first attested in stories from the Middle Ages. They are almost always small and grotesque, mischievous, or outright evil.

Therefore, the leitmotif of all these horror movies is that the food chain reverses, and nature does exactly what we do to nature: we come to a tree and we cut it.

MANOUCHEHR: Again, there is great empirical evidence on what kind of cultural phenomenon comes with a certain economic cycle.

For example, zombie movies are connected to a different economic development than vampire movies. The vampire stands for an old aristocratic person. Right now, there are new vampire movies coming out, but there are no zombie movies. You don't see many zombie movies today.

You could say that we are now in the vampire era, because people are looking for ancient leadership. That's what vampires represent. Zombies don't have leadership. But they are a group of creatures with a very rational choice and momentum.

TOMÀŠ: The nice thing about zombies is that every adult has a basic dilemma: Do I reproduce or do I eat? This is the dilemma that you're solving every second in your head. Zombies are so efficient, because they reproduce by eating.

> We're in the vampire era.
People are looking for
ancient leadership. <

3
How Digital Activism is Changing the World

Geraldine de Bastion
Raúl Krauthausen

PANEL DISCUSSION, BERLIN, MARCH 21, 2015

Digital innovation has strengthened existing networks and enabled the development of new ones. Through digital tools, people are communicating, creating strategies and projects for social change. From fighting corruption to finding lost children to raising awareness for illnesses and catastrophes around the world, Internet, particularly mobile Internet, has brought people together to tackle social challenges.

Equipped with the right tools and an appropriate understanding, individuals and groups of likeminded people can apply agenda settings as easily as any NGO and lobbyist. Even fundraising and organizing volunteers can be managed digitally. Starting a grassroots campaign, organizing knowledge through crowdsourcing, or securing financing through crowdfunding, is just a click away.

How can these innovations drive social change on the local and global level, and how do theses digital efforts translate into actual impact? Which media channels are defining our perception of reality, and who are the new thought leaders?

Activist Raúl Krauthausen and digital communications expert Geraldine de Bastion discuss the current developments in digital innovation for social change. Both Raúl and Geraldine are catalysts in the European NGO digital scene, lending us clues on how we can influence and become a part of social change.

PANELLISTS

GERALDINE DE BASTION is a freelance international consultant with a multicultural background based in Berlin, Germany. She is an expert on information and communication technology and new media for development. Using this expertise, she advises governmental organisations, NGOs and businesses on digital media and communication strategies. She also works with activists and bloggers around the world. Geraldine began her career working for the Gesellschaft für Internationale Zusammenarbeit (GIZ) and the German Federal Ministry for Economic Cooperation and Development (BMZ). She helped organise the German Delegation for the World Summit of Information Society (WSIS) in 2005 and worked for the Philippine National Development Agency (NEDA) on behalf of GIZ. During the past few years, Geraldine worked with newthinking communications GmbH, an agency for Open Source strategies. In 2012, she curated re:publica, Germany's largest conference on Internet and Society. In her free time, she is also a member of the non-profit organisations Digitale Gesellschaft e.V. and ICE Bauhaus.

RAÚL KRAUTHAUSEN is an activist and social entrepreneur. He initiated the crowdmapping project wheelmap.org, which collects and organises information about wheelchair accessibility throughout cities. With "Sozialhelden" (Social Heroes), he created an NGO for social campaigning and change. Raúl is a prominent voice amongst international social innovators. He is a recipient of the Order of Merit of Germany and an Ashoka fellow. Peruvian-born, Raúl Krauthausen resides in Berlin and travels to the furthest parts of the world to engage with local communities.

Geraldine, you work as a consultant with newthinking[70], as an event creator with re:publica[71] and also as an activist with Digitale Gesellschaft[72].

What are your favourite tech-related social projects? Which of them really made a social impact in the last few years?

> GERALDINE: Let's begin with Ushahidi[73]. It's still one of the best examples of a really powerful platform: a tool that was created to foster social change.
>
> Ushahidi is a crowd-mapping tool that was created after violence broke out in Kenya following the national elections in 2007 and 2008. It was created by the very vibrant blogger community, because there was a media blackout, there was a lot of chaos, and they wanted to create more oversight and give the people a bit of power and confidence. So they created a mash-up tool between Google maps and a tool called FrontlineSMS[74] that allows you to input information both via the net and SMS short code.
>
> Through this, they started mapping all the incidences of violence and encouraging citizens to be good examples with non-violent behaviour. It began as an open-source platform, and after that, Ushahidi grew into a portal that

70 newthinking is a company for open source strategies and projects, working with the diverse interfaces of new technologies and societies.

71 re:publica is a conference in Europe that deals with the Web 2.0, especially blogs, social media, and information society. It takes place annually in Berlin.

72 Digitale Gesellschaft is a German registered association founded in 2010 that is committed to civil rights an consumer protection in terms of Internet policy.

73 Ushahidi, Inc. is a non-profit software company that develops free and open-source software for information collection, visualisation, and interactive mapping.

74 FrontlineSMS is a free open source software used by a variety of organizations to distribute and collection information via text messages. It can work without Internet connection and with only a cell phone and computer.

enables people to host their own crowd maps around the world.

In cases of humanitarian disaster, it's one of the main tools used today. For instance, it was used after the tsunami and devastation in Japan; official help organizations used it to identify victims after the earthquake hit Haiti; and it's been used for different political cases around the globe.

Here are two examples of how the platform has been used in Kenya since its creation.

One is Uchaguzi[75], which is a platform that was created for election monitoring both in Kenya and in other African countries, as well. It's used successfully to allow citizens to report back from their election areas and report on the process of the elections. They can say whether there were any funny things going on or any transparencies. It's been recognized by official election monitors as a great tool to foster citizen participation and to complete the picture. They can't go everywhere.

Hate speech was one of the main factors that caused this violent outbreak in 2007 and 2008. So the team around the iHub[76], which is an innovation Hub in Nairobi, started a research project called Umati, another instance of Ushahidi in Kenya, which monitors hate speech. It's the largest project about hate speech

75   Uchaguzi is a joint initiative spearheaded by Ushahidi in 2010 to monitor the Kenyan constitutional referendum. Ushahidi collaborated with the Hivos Foundation, the Constitution and Reform Educational Consortium (CRECO), INFONET, and the Canadian International Development Agency to develop an ICT platform, which enables citizens, the civil society, election observers, law enforcement agencies and humanitarian response agencies to monitor election incidents of significance using any technology at their disposal.

76   iHub serves the tech community by connecting organizations and people, building market relevant solutions, and being ahead of the curve of innovation. It aims to become the best African support system for ICT focused tech entrepreneurs.

globally. It's recorded all different kinds of instances of hate speech for years.

This is a really significant research project for mapping these things but also to spur discussion on how to deal with hate speech online. Perhaps, rather than cracking down with government regulations, they can find ways for communities to directly address and work against hate speech.

Ushahidi is one of my favourite examples for an open-source platform that enabled many different ways of social change across the world, but especially in the country where it was created.

I wanted to ask you, Raúl, about platforms that have been in the public eye through the media, things like Campact[77] and Change.org[78]. What are the advantages and disadvantages of these sorts of engagement platforms?

RAÚL: These platforms give every one of us the opportunity to raise our voices for special topics. The danger is that we face too many possibilities to vote for or against something, and sooner or later, there will be a flood of petitions voting for or against something. After that, nobody will do it anymore, or no politician will pay attention to it anymore, because everyone is doing it for all the cases. For example, they might say, "I don't want Markus Lanz[79] anymore at *"Wetten, dass ...?"*[80]" Or it might involve a really relevant social political topic like

77 Campact is a non-profit, non-governmental organization established in 2004 and based in Verden. It offers an Internet-based participation forum.

78 Change.org is a petition website operated by for-profit Change.org, Inc., an American certified B corporation, which claims to have over 100 million users. It hosts sponsored campaigns for organizations.

79 Markus Lanz is a presenter and producer on German television, having hosted Markus Lanz on ZDF since 2008.

80 Wetten, dass..? is German for "Wanna bet, that..?" It's the long-running German-language entertainment television show and the most successful Saturday television show in Europe.

transparency or the end of nuclear power or whatever, which might me more relevant.

As I told you before, there was a petition on Change.org dealing with a boy with down syndrome. Today is "World Down Syndrome Day," because today is the 21st of March. So there was a petition at Change.org created by the parents of this boy who wanted to ask for help so that their son could go to the regular gymnasium.

The petition got a lot of media attraction here in Germany, but there was also a counter petition on Change.org to not let this boy go to the regular gymnasium. That was tragic, because the person who suffers is this child, and if you, as a child, face a petition against you, this might be a very hard and sad situation for the parents. It's very emotional.

I don't see the political dimension behind being against a boy coming to a gymnasium. The political dimension might be rather about how can we support schools to be more inclusive, but surely not on we *don't want this boy in our class.*

That is a big challenge we have to face when we deal with Change.org or other petition platforms. They open up great diversity of political dimensions.

> Online petition
platforms open up
a great diversity of
political dimensions. <

What is the advantage of Campact? We also discussed that they're doing this a little bit differently.

RAÚL: In comparison to Change.org, there's Campact. What I like about Campact is that they focus on special topics. There's an office in the background that research relevant topics to try to really make action to change the decision maker's opinions. They research how we can influence them to decide the other way. This is something that can be moderated, curated in a better political way than Change.org, where everyone can start a petition.

I don't believe in giving everyone the possibility to vote for everything, because you might need more information, instead of yes or no.

Raúl, you are very experienced in doing campaigns off and online. What do you think are the biggest challenges to bring these two dimensions together?

RAÚL: My most-liked, non-successful campaign was the KONY campaign[81], which got a lot of media attraction. But nothing happened. Everyone was meant to talk to his or her politicians by the 20th of April, but nobody did it and nobody got the message.

What I believe is that it's very important to focus on one special topic, like something you can really experience in your daily life. For example, this is found with our project wheelmap.org[82], which is just an online crowdsourcing platform where people can rate restaurants and cafes by accessibility for people with a disability. It has the question: does the entrance have steps in the front—yes or no? That's the only question we ask. And that is something everyone can understand. But nobody runs through the city with a ruler to measure out how wide the door entrance is or how high the step really is.

81   The KONY campaign was a short documentary film produced by Invisible Children, Inc. It was released in March, 2012, and its purpose was to promote the charity's "stop Kony" movement to make Ugandan cult and militia leader and indicted war criminal Joseph Kony known in order to have him arrested by the end of 2012, when the campaign expired.

82   wheelmap.org is an online, worldwide map for finding and marking wheelchair accessible places, developed by the German non-profit organisation Sozialhelden e.V. Anyone can find and add public places and ate them according to a simple traffic light system.

So, regarding your question. Balancing online and offline activities is based on the question of what can be done online, and how precisely you ask your user. It's also based on what kind of information you want from the user. Sometimes, it's just easier to ask for either yes or no. But the user won't add a postal code or add the height of a step, because the user doesn't know how high it is.

Of course, wheelmap.org is only a mapping platform, where we can see which place is accessible or not. Until now, we haven't made any place accessible. So the political dimension behind this platform might be: how can we raise awareness for restaurant owners or politicians to make laws or to buy ramps for cafés and restaurants to make them more accessible?

That was the reason we started the second project, called wheelramp.de, where you can just buy a ramp. It's a single-service website, a one-pager. Just buy a ramp, 1.2 meters long. We don't allow you any excuses anymore. You know where to buy a ramp.

Geraldine, many campaigns are also campaigns that gather likes and shares. What is your take on these campaigns that are very prominent on social networks?

GERALDINE: The net has so much space for so many different kinds of forms of participation. Of course, there are things that are just trends, and we wonder about their real impact and the message they have.

The most prominent the last year has been the Ice Bucket Challenge[83]. You could say that this campaign did spur debate about a disease that otherwise nobody

83 *The Ice Bucket Challenge, sometimes called the ALS Ice Bucket Challenge, is an activity involving the dumping of a bucket of ice and water over a person's head to promote awareness of the disease amyotrophic lateral sclerosis and encouraging donations to research. It went viral on social media during July of 2014.*

would have known about. However, very often with these campaigns, like KONY, there's very little follow-up and there's very little communication to the participants about what actually comes after.

When you organize participatory projects, you should always have this rule of thumb, saying you have to involve the people—not just in the height of the phase of the action, but also in the outcomes. You should communicate what came of it. That's something that those campaigns definitely lack.

> When you organize
participatory projects, you
should always involve the people
not just in the high phase,
but also in the outcomes. <

Sometimes, in activism, it's just important to show that these many thousand people support a particular topic and to feel that there's a massive body of people backing something. This way, you can open doors to perhaps talk to politicians and have something in your hand to show.

I think those campaigns can only really be taken seriously if they are accompanied by other forms of political action.

20 years ago, this was all a bit different. When Raúl started his first NGO Sozialhelden[84], he was in his mid-20s. It was the beginning of the Internet. How has your work changed since then? Is it basically the same, with a little bit more of a digital dimension?

84   The Sozialhelden is a registered non-profit association with headquarters in Berlin. It organises a network of volunteers who are committed to various social justice actions.

RAÚL: When we started in 2004, we already used the Internet as well as we could, because we didn't have an office. So, we managed a lot by email or by a customer relationship management tool called Highrise, from 37signals, which is what we still use today. We started to manage everything online, cloud-based, and later on, there came services like calendars, online sharing and things like Dropbox or BTSync. We are still using many cloud-based office management tools often used by non-profit organization, but now we have an office. That's the only difference.

With an office, you have the advantage of regular work times. You come at 10:00 and leave at 6:00. Before that, we just had the situation that we met as a team in a café or in a restaurant, my apartment or at IKEA or even in the Ringbahn[85]. Interestingly, IKEA is actually a very comfortable place, if you want to start a start-up, because you have office space, you can buy everything, you have power supply under the tables, you have a cantina, and you have toilets. You're only lacking in Internet. But today, with a UMTS stick, everything might be easier. We worked there for two days, and it worked very well.

> Now, you can use
wheelmap.org in Klingon.
I don't know if it's necessary
or useful, but it works. <

What I really like in an online world is that you have a community you can work with. For example, we provide wheelmap.org, in 23 languages now. Of course, none of us is able to speak 23 languages. We started in German and English, and then we asked the wheelmap.org community what kinds of language they could add to the wheelmap.org as users.

85   The Ringbahn is a Public Transport System in Berlin.

How Digital Activism is Changing the World

We gave them an online possibility to translate every single word we use on wheelmap.org. This way, the fifth language online was Klingon, because there was someone out there, I don't know who it was, who could speak Klingon. Now, you can use wheelmap.org in Klingon. I don't know if it's necessary or useful, but it works.

The sixth language was Japanese. Now, we have Polish and many other languages. That is something you could only do via the Internet: ask users for help and be transparent about what help you need.

GERALDINE: Betahaus[86] is probably very happy that IKEA didn't have Internet, otherwise they wouldn't have a business model today.

I was just thinking: back when the Internet was young—this is great, this reminiscing about the Internet—in 1999, when I started university, we started an NGO, which connected different youth magazines and student papers. For me, one of the big differences back then was that we had to create our own tools.

We built a kind of blogging platform, because there was no blogging platform service to allow young people to publish and tell their own story and write autonomously. That was a very adventurous time, because you had to have way more technical capacity than you need today. Another difference that I would point out is: back then, the Internet was rougher, and the possibilities of what you could do politically were also much more experimental.

> Now that the Internet has matured, there's much more legislation on what is legal and what's not. <

86   Betahaus is a space to create, work together, and re-imagine the future of work—eliminating hierarchies and old structures along the way. They have locations in Berlin, Hamburg, Barcelona, and Sofia.

Back then, we had very passionate discussions about using DoS Attacks[87] or Denial of Service Attacks, as a legitimate form of protest, like a form of demonstrating on the net. Back then, you needed a lot of people to do a Denial of Service Attack, not one automated system, like today.

Now that the Internet has matured, along with the tools that we use, there's also much more legislation on what is legal and what's not. Some of the tools that we've discussed have been integrated into our political systems, like petition platforms, for instance.

Geraldine, through your curating work with re:publica, you got in contact with tech start-ups from Africa, which made you produce a documentary movie on the tech scene in Africa, called *Made in Africa*. What was the main message of the film, and what were the reactions?

GERALDINE: This film has been quite a journey. When I'm not organizing re:publica or doing events or digital activism in Germany, my actual day job lies in development of corporation and working with digital media companies mainly in sub-Saharan Africa.

Over the years, I've gotten to know so many exciting and amazing people that do such great work to build up tech communities in their countries, to really try to create social and economic change, without waiting for big donors or big organizations. They're just taking matters into their own hands.

Through some of my work with re:publica, I'm trying to make the discussion more global. In Germany, we have a tendency to, as we say, "swim a bit in our own soup." We see a lot of the debates that we lead from a very national point of view.

87 *The Denial-of-service attack is a cyber-attack where the perpetrator seeks to make a machine or network resource unavailable to its intended users by temporarily or indefinitely disrupting services of a host connected to the Internet.*

For instance, the debate we've had about quality journalism versus bloggers for many years is led from our perspective, which we have because of our very high standards in journalism. It's sometimes important to break that open and see things from a more global perspective.

> I wanted to show a modern, realistic version of life in Africa, its digital innovation, and the stories I've come to care about. <

I started bringing in more and more colleagues from other continents to speak at re:publica. In 2013, we hosted something called the Global Innovation Gathering[88] and brought together innovators from across the world to come and speak at re:publica. A friend of mine was doing some filming work there, and he said to me at the end of the conference, "Geraldine, this is cool. You're getting more people to know about it. But this is not enough. Why don't we make a movie about it, so more people will know about the great stuff happening in Africa?"

I'm not a filmmaker, so it was a big experiment. But for me, the main aim of making this film was to create some piece of information that would show a modern, realistic version of life in Africa; to share some information about its digital innovation and the people's stories that I've come to care about over the years. I wanted to break some of those clichés that we see through media. The film doesn't wipe away any of the problems or critical issues, but it's a film about happy, well-adjusted people doing cool stuff in Africa.

88   The Global Innovation Gathering is a group of fast-growing innovation hub founders, community managers, makers and hackers from across the world, who exchange ideas and collaborate online and offline at conferences worldwide.

What happened when the people there saw the film?

GERALDINE: That was a really overwhelming experience for me. This was last year at re:publica. When we showed the rough cut of the film for the first time, I got to moderate Stage One, and I had never been so nervous at any re:publica. I showed the movie for the first time, with all its protagonists there. It was overwhelmingly positive and really emotional for me, too.

The film brought up a completely different use case, as well, which was not just showing it to the people at re:publica for a better understanding, but to show it also to the politicians and people in the protagonists countries. They need to see it, too, to create a better understanding for the work they do. We have to translate a lot about the work we do for our parents, or people working outside of our bubble. They felt the film was a great tool for this translation.

How would you assess the use of the Internet as a tool for big social movement? Looking at it internationally, I think a global average of 40% have real access to the Internet. How would you reflect that to really use digital media for bigger movements?

GERALDINE: Of course, there are still huge issues about connectivity and access. Those are not just being put aside. The majority of the world's population still lacks Internet access, especially any kind of broadband Internet. However, I think there are many great examples that show us that you can use different kinds of cross-media approaches to overcome those barriers.

There are projects like U-Report[89] in Uganda, which is an SMS telephone-based initiative that connects with

89 U-Report is a free SMS social monitoring tool for community participation, designed to address issues that people care about. Real-time response information is collected and the results are shared back with the community.

young people across the whole country to do different kinds of opinion polls and ask young people questions that are then published in other kinds of media channels, like radio shows and newspapers. The polls are also brought forth to politicians to make them see what young people think.

Look at the possibilities you have when you are digitally connected, not just via smartphones, but also feature phones, and with traditional media. I do believe that basic technical tools can have a very powerful impact, even in developing countries.

Raúl, what strikes me with your project is that it's often local, hands-on, real life. As you said, wheelmap.org is based on one challenge, and you addressed it. Most recently, you started a project called brokenlifts.org. Maybe you can tell us a little more about how that came about and its purpose.

RAÚL: A lot of people asked us if we could do something about navigation for people in wheelchairs. How can I get from A to B? Because, through wheelmap.org, we already know which places are accessible. We know B might be accessible, but how can we get there?

Routing is a very complex problem. We found out that mobility, for people with mobility impairments, really relies on working elevators or lifts, especially for places with public transports like Berlin.

We found out that the BVG, the Berliner Verkehrsgesellschaft, published a hidden RSS feed of their not-working elevators, but the S-Bahn [90] didn't. We started scraping the website of the S-Bahn, where they showed the broken elevators, and merged it with the BVG data. This way, we could have one platform, which shows us which elevators of Berlin-Brandenburg public

90   *S-Bahn is a public transport system in Berlin and part of the Verkehrsverbund Berlin Brandenburg (VBB)*

transport systems are working or not, given by the transport organizations.

What nobody knows is that elevators are online. They tell the BVG automatically if they are working or not. They have SMS modules inside, and they just send a ping. It's not a crowdsourcing platform; it's not something where people can tell us if an elevator is broken or not. This information wouldn't be reliable for long.

Now, we're talking with Deutsche Bahn to scale this project to all of Germany, but it's a very complex system.

You are supplying or creating a public service. How is that being compensated? In what way have you worked with the officials?

RAÚL: We started with a hackathon at the Random Hack of Kindness[91] three years ago. That was the moment when we started scraping the data from the S-Bahn illegally. Two months later, the Berlin minister of economy, tech and science, the minister of *Wirtschaft*, *Technik* and *Forschung*, said to us that open data is a nice topic, because we want to open this data to the public. Berlin wants to be one of the most active cities in dealing with open data.

> Berlin wants
to be one of the most
active cities in dealing
with open data. <

They asked us if we could do something with Verkehrsverbund Berlin Brandenburg, which is the head organization of BVG and S-Bahn. They paid us to make

91   *Random Hacks of Kindness is a community of technologists who solve problems for purpose-driven organisations, holding two hackathons a year for social good.*

this project happen. The idea was to develop it, so that you can get an SMS or an email if an elevator you chose is broken. However, right now, we are just showing which elevators are running or not.

In Berlin, we have 400 elevators, more or less. 16 of them are broken right now. This could become a good tool to discuss how we can make the service better. Maybe we can gather information from these occurring, broken situations.

That's interesting, finding hidden patterns. That's great. Also maybe that's a question for both of you.

There used to be websites on how to fix your street on fixmystreet.com. In Britain, it was quite successful. Do you see potential for city governments to utilize these sources through crowdsourcing to get rid of challenges like these?

GERALDINE: I definitely see the potential of involving citizens. I think the projects that Raúl does are such great examples for how to empower citizens by giving them tools to do that: to contribute to a functioning environment in which they live. The examples that you stated are also great platforms. They've been copied and implemented in other countries, as well.

The ambivalence, perhaps, or the danger begins when the government increasingly outsources those services to citizens and thereby rids themselves of the responsibilities. It's their responsibility to maintain functioning infrastructure and a functioning environment for citizens.

It's great when there's a positive collaboration between government and citizens, involving reallocation of power and information. But they shouldn't serve as an excuse to privatize government responsibilities.

> The danger begins when
the government increasingly
outsources those services to citizens,
thereby riding themselves of
the responsibilities. <

RAÚL: I would totally agree. We found out last year that there have been several projects in Berlin that wanted funding from the Berlin government. They were told you can only get money if you work together with wheelmap.org. But we as wheelmap.org have never seen any money from the Berlin government. That is an outsourcing situation.

The wheelmap.org platform costs us a quarter-million a year, so it's a very big advanced project. We have an Android app, an iPhone App, which is very expensive to maintain and make updates on. Most of the costs are personnel costs, so it's not a big office or a yard. It's just people.

This makes me angry. We are running this project, and the people expect to have this project forever. They ask us if we can implement new features. Apple offers a new iPhone with a bigger screen, so we have to make updates, but nobody pays us. We have to start funding and fundraising.

When we ask the government for money, they say, "You've been running for four years. Why did you fail?"

We said, "We aren't as innovative as we might have been four years ago, but we need some kind of infrastructure support." In Germany, you just can't get it.

How can you avoid that? How can you avoid outsourcing?

RAÚL: The government wanted us to work with companies like ImmobilienScout24. We worked with Google.

But we are still looking for money. We are still making crowdsourcing campaigns and asking for donations. It's working, kind of, but we are still looking for money.

The idea behind wheelmap.org is not to become the biggest project in the world. We don't want to compete with Google maps. Our goal is to raise awareness for the topic of accessibility for places like restaurants and cafés or discos or cinemas. We want to convince Google, we want to convince Yelp, and we want to convince Apple maps to implement our information in their databases. That's the reason we opened our data. We are based on OpenStreetMap.

All the information we gather is synchronized, every minute, with the OpenStreetMap database. OpenStreetMap is like Wikipedia on a map. That's what makes wheelmap.org sustainable. If we die tomorrow, the information would still maintain on OpenStreetMap, because we can't guarantee working for the next 50 years on this project. It's really expensive.

How would you measure your success as an NGO? It's obviously not revenue.

RAÚL: A founding friend of mine said that when we have one million spots on wheelmap.org, he will quit. That means it's okay; it's working. Right now, we have 500,000 spots on it, after three years, so it's growing well. I assume that in three years, he will quit.

> We believe we can implement the idea of disability mainstreaming into other companies, like advocating, but on a very modern level. <

We call our mission *disability mainstreaming*. We want to raise awareness on accessibility and disability topics for people without disabilities. If you can find out on Google maps if the place where you want to go is accessible, that is the most mainstream we can achieve with the information we have.

We are working together with companies like ImmobilienScout24 on finding accessible flats for people with disabilities. Before, there wasn't a platform Germany-wide where you could find accessible flats for your grandmother. We started by implementing a new filter option in that database. Now, we're the biggest platform dealing with accessible flats.

As a non-profit, we believe we can implement the idea of disability mainstreaming into other companies, instead of doing everything ourselves. It's more like an agency, more like advocating, but on a very modern level, using smartphones, using modern tools and technology.

Over the years, you started utilizing different tools. Social media hype, as such, is over, becoming a normality. It's part of our everyday communications.

When you plan your activities, then, how do you involve social media? How do you use other social networks in your NGO activities, such as Instagram, Pinterest? What's the banquet of your everyday work?

RAÚL: All of our projects have Twitter and Facebook accounts or pages. The most successful account is the Facebook account of Leidmedien.de. It's a website on which we want to teach classical journalist from *Spiegel Online* to *ZDF* how to speak about people with disabilities without pity, without raising bad or heroic emotions.

Not everyone who has a disability is suffering, and not everyone who has a disability is a hero. Most of the journalists write like that, however, using phrases like "… despite his disabilities, he's surviving…"

We started this platform to educate journalists on how to write politically correct. We started two weeks before the Paralympics 2012 started; we call it Fettnäpfchenzeit[92]. My colleagues started the Facebook page, and we discussed articles we found on the web about people with disabilities with our users.

Through these discussions, we got an idea of what the biggest problems are for journalists. Most of the time, journalists don't have regular personal contact with people with disabilities, until the moment they have an interview with them. The interviewee might be the first person with a disability the journalist was confronted with in his whole life.

Seeing these discussions on Facebook, we found out that journalists struggle to find the right words. For example, are they allowed to say *Mongo*[93] to a person with down syndrome? I still read that term in newspaper articles. Or are they allowed to say *cripple*? Or how do they find a good picture of someone in a wheelchair?

There is one particular picture of a girl in a wheelchair, in which the wheels of the wheelchair are twice as big as the girl, because it has been photographed from bottom up. The wheels seem very big. This very picture is always used in newspapers, on *Tagesschau*[94] and *ZDF heutejournal*[95], when they talk about inclusion and

---

92   German term, meaning "time to drop a clanger".

93   Mongo is short for Mongole, which is an extremely rude word used to refer to someone who has down syndrome.

94   Tagesschau is a news broadcast of the ARD, which is produced by ARD-aktuell and is shown daily in Das Erste and as live stream on tagesschau.de. It's the oldest existing news program on German television.

95   The heute-journal is a news magazine of ZDF, dealing with individual topics of the day in more detail, offering background information.

education. The picture shows the main problem: the wheelchair with its big wheels, which are the big life obstacle. It's disturbing. But it does not show the educational part, the inclusive part at school, for example, where people are learning together.

What we do now is start a picture database with good images of people with disabilities. These will be good images for journalist to use for *Tagesschau* or *Tagesspiegel*. This way, they can replace and delete the bad images from the dpa[96] image databases or Getty[97].

GERALDINE: I want to ask the question from a different perspective. You have so many great examples of using tools that are so easily available. We spoke about the history of Internet tools in our early projects. How easy is it to keep up with new platforms and new tools and integrating those not just privately but also in your work?

As an example, a couple of weeks ago, we constantly talked about Facebook soon coming to an end, that some new platform will arise. We learned this time is coming soon. Cabinet of Germany[98] joined Facebook mid-February this year. Did anybody visit the official Facebook page recently launched by the Cabinet of Germany?

RAÚL: It's better than expected.

96   The dpa is the German Press Agency, or the largest news agency of the Federal Republic of Germany, with headquarters in Hamburg and central editorial offices in Berlin. It's represented in approximately 100 countries worldwide.

97   Getty Images, Inc. is an American television agency founded in 1995 by billionaire Mark Getty and Jonathan Klein. Their headquarters are located in Seattle, London, and the German subsidiary in Munich. It has archived over 80 million images and illustrations, along with 30,000 hours of film footage.

98   The Cabinet of Germany is the chief executive body of the Federal Republic of Germany. It consists of the Chancellor and the cabinet ministers.

GERALDINE: You think so?

RAÚL: Yeah.

GERALDINE: I had such fun. I laughed so hard. I recommend visiting it purely for entertainment purposes, but also because the Cabinet of Germany decided to add a post for every major historical event and every chancellor dating back to Konrad Adenauer[99]. You keep scrolling down wondering what's next. It becomes weirder and weirder by the minute. They obviously decided that the Internet is not complete if the Cabinet of Germany doesn't input Germany's entire history on the Facebook page of the Cabinet of Germany. It's done now. What a relief.

RAÚL: Within two weeks.

GERALDINE: Within two weeks. I visited a panel at one of my favourite events last year, at Reeperbahn Festival[100]. It was a panel of just teenagers from 14 to 16 years old, explaining the Internet to grownups.

This was a really fascinating session, because they said things like, "I'm only on Facebook to talk to my teachers and parents. I don't actually spend any social time there." It was also very insightful when it comes to what kinds of tool they use. For instance, there was one strong case for using Snapchat[101] rather than

99   Konrad Hermann Joseph Adenauer was a German statesman who served as the first post-war Chancellor of the Federal republic of Germany (West Germany) from 1949 to 1963. He led the country from the ruins of World War II to a productive and prosperous nation.

100   The Reeperbahn Festival is Europe's biggest music and tech festival, held each year in late September. In 2017, the 800th Reeperbahn Festival featured 800 programs, more than 37,000 specialists, in more than 70 venues in St. Pauli.

101   Snapchat is an image messaging and multimedia mobile application created by Evan Spiegel, Bobby Murphy, and Reggie Brown, former students at Stanford University. The pictures and messages are only available for a limited amount of time before they become inaccessible.

Instagram[102], because of the durability of the pictures. Instead of staging your life and having the perfect selfie, you'd rather use something like Snapchat, because you don't have to make your life look perfect all the time. It's just visible for the moment.

I'm really comfortable using all the tools that we're accustomed to in my work. I wouldn't know how else to do my work, without them. But to be honest, I don't find it the easiest thing to keep up with all new platforms.

RAÚL: It's very hard. Last week we started our first WhatsApp[103] channel, just to test if there's a younger target group. This year, I really want to start push notifications. If the user allows it, we can even target people by regions via our iPhone app.

We can tell every person with the wheelmap.org app in Hamburg or in Berlin to make a run on something. That's how we could activate people again. It's not very interactive. It's just one-way push notification. You can't push back a message as a user. But for now, this is what we're working on for the future.

If you look at campaigning, for instance, there are many different types of campaigns. Some are more political, others less, like the Free Hugs campaign or the Ice Bucket Challenge.

How do you feel about social networks and campaigning if they're often more "entertaining" in nature? How serious can a campaign be on a social online platform?

102   *Instagram is a mobile, desktop, and Internet-based photo-sharing application and service. It was created by Kevin Systrom and Mike Krieger and launched in October 2010 as a free mobile app on iOS. It allows users to apply digital filters to their images and add locations through geotags.*

103   *Whatsapp Messenger is a freeware and cross-platform instant messaging service for smartphones. It uses the Internet to make voice calls, one to one video calls, send text messages, images GIF, etc. using cellular mobile numbers. All data are end-to-end encrypted.*

RAÚL: Today, I went on YouNow[104] for the first time. YouNow is a live streaming video platform. You login, and you're online, if you want to be. You stream your life via webcam picture. It's like Google Hangouts, but everyone is watching.

GERALDINE: As bad as Chatroulette[105]?

RAÚL: Chatroulette is random, but with YouNow, you can go to a specific person and stay there. The most awkward moment I experienced was when I visited the channel of a famous person. The girl's only goal was to get more than 5000 likes. If she got 5000 likes, she was going to eat a lemon.

I watched for 10 minutes. She took her laptop, went through her house to the kitchen, got a lemon, showed her rabbit, her parents, her sister and so on, and then went back to her room. Within these 10 minutes, she got 2000 new followers. Not likes, but followers.

I was so blown away by how fast this community grew while the quality of the conversations was so bad. 200 people were chatting to one girl, who of course couldn't answer everyone.

GERALDINE: Just because of a lemon?

RAÚL: Most of the conversations were about fucking. They all wanted to fuck this girl, and that was very awkward.

GERALDINE: Yes, that is very awkward. But we talk about this so much. Usually, the conclusion is that societies

---

104 *YouNow is a live streamed video chat service, where users stream their own live video content or interact with the video streams of other users in real time.*

105 *Chatroulette is an online chat website that pairs random users together for webcam-based conversations. Visitors to the website begin an online chat with another visitor. At any point, either user may leave the current chat by initiating another random connection.*

meet online. The level of conversation is probably pretty banal most of the time. Yes, these sites can be used for great, very profound things, but they can also be used for really random things.

I guess crowdfunding is another very good example. I failed dismally trying to crowdfund for my film. I thought I had a really good cause with a really well explained, very clear mission. I had a really low target, and I have a relatively good community. I thought, "This is going to totally happen; this is going to work out." But it was just terrible. It was a horrible experience.

Which platform did you use?

GERALDINE: Indigogo[106]. I won't say it was the platform's fault at all. It was a lot of different factors. Then, at the same time, you watch somebody who's making potato salad get $500,000. You think, "Wait a minute. There is something broken with this."

> You watch someone who's making potato salad get $500,000. You think, "Wait a minute. There is something broken with this system." <

People want to be entertained. People are willing to click to like something, click to share something, click to fund something that is just entertaining.

RAÚL: When we talk about these new social media technologies, I sometimes have the feeling that we believe

---

106   Indiegogo is an international crowdfunding website founded in 2008 by Danae Ringelmann, Slava Rubin, and Eric Schell. Its headquarters are in San Francisco, California. It was one of the first sites to offer crowd funding.

that the past was better, but it wasn't. There was bad content, too.

Maybe we need to take some time to develop proper storytelling formats for YouNow, so that it's for a cause, rather than something about your best beauty product.

GERALDINE: Remember how people spoke about Twitter when it first started. They said something like, "Twitter is the toilet door of the Internet." Now it's one of the most important tools for journalists online.

RAÚL: Yes, now you can really make a lot of traction through Twitter for your cause, especially with a hashtag.

Jan Böhmermann[107] is making a hashtag every week and becoming a top trend.

You remember the One Laptop Per Child project by Nicholas Negroponte and The Hole in The Wall project by Sugata Mitra. Negroponte says that devices are important to get access or to create a social change. Mitra claims that it's rather the information and how you share it that's important.

What do you think: is it the device, or is it the information to start a movement?

GERALDINE: It's a tricky question, especially in developing context. I want to give a bit of a broader answer. Mitra's assumption is based on the fact that he is a physical scientist. He developed this mathematical calculation on how children learn best, not as individuals, but in small groups. He did this by supplying technology for children to learn, based on those mathematic and scientific assumptions about giving people a shared device to use in a group and using it to communicate with one another.

107   Jan Böhmermann is a German satirist and television presenter.
He has also worked as a comedy writer and producer.

Both projects are very interesting to watch. Many different cases around the world show various, successful ways children can learn with technology. We discuss often whether information or access has to come first. That discussion is currently centred on zero rating services in the world. It's a very fundamental discussion when it comes to the question of net neutrality and what kind of access should be provided to people.

For instance, Facebook first launched their service Facebook Zero in many countries across Africa, and now also in the whole world. That means they have a deal with the ISP[108], the telecommunication providers. Everybody who has a phone in Africa, often not a smartphone but a feature phone, can access Facebook for free, independent of whether they have a data contract. It doesn't cost people to access Facebook.

>Facebook is swallowing the Internet.<

For many people in developing countries, Facebook is the Internet, because they can access almost everything through Facebook. They can communicate with people, but they cannot go beyond that. Now there are different initiatives picking up on that idea. Facebook has started the initiative internet.org, offering a number of other services, including *BBC*.

You could say this is great, because people need information first. But I find it really scary, personally, because it gives you a pre-curated version of the Internet. It feels like Facebook is swallowing the Internet. Whilst the platform is becoming increasing unimportant to young people as a social media platform in our countries, Facebook is presenting itself as the accessible version of the Internet where people cannot otherwise access.

This is a very tricky debate. In many countries, you could say the information is the most important thing. Would you not want, for instance, for people to have access to Wikipedia in Ebola-stricken countries? They could inform themselves about the disease and the consequences.

Looking at the greater effects of this and what it implies for the Internet and Internet freedom, I think it would be great if these big Internet platforms and companies would fund a minimal amount of free Internet in collaboration with ISPs. Then, people could choose to go to whatever site they want, rather than going to a pre-curated site.

Therefore, access matters, perhaps more than information.

Let's come back to the entertainment factor of campaigns. You chose a book as a format to tell your story. The biography is out now, called *I Didn't Want to be a Roofer Anyway.*

The thing about a book is there's a beginning and there is an end. This book also has the sad and the funny aspects of your life and your perspective. It's a great read. How do you continue to tell your story digitally? How do you use humour to make people better understand you?

RAÚL: When the publisher asked me if I wanted to write a book, I said no, because they wanted me to write a biography. I said, "I'm 33 years old. I don't want to die tomorrow. So it would be only one half of a biography." It was a weird question.

Then a friend told me that not everyone is being asked by a publisher to write a book, especially Rowohlt Verlag. Normally the situation is the other way around: that you are looking for a publisher to publish your book. This was a chance I had to take.

> I'm still in the process of accepting my disability. I'm not done yet. <

Of course, I'm a blogger. I'm a Twitterer. I'm a Facebook user, but I'm not a writer. I went to the publisher and told them that I actually I hate writing. An empty white page frightens me.

As a reaction, they offered me someone to write a book like a co-author. That was Marion Appelt. Together, we started to think about what could we possibly write.

In the end, it naturally became a biography, but it does not start with my birth and go until the present. It just has a different point of view. It's my journey on how I started accepting my own disability. That's actually the plot of the book and something I'm still dealing with. I'm still in the process of accepting my disability. I'm not done yet.

I believe when I die, I will still not accept it with all its situations and moments. But in writing the book, I also started to realize that there is a big political dimension on why it is so hard for me and for others to accept disabilities, both someone else's disability and their own.

Maybe you can compare it with gender topics. Even if 50% of society is women, women don't have the same possibilities to fight for their rights that men have. The same situation is happening with people with disabilities.

Six years ago, I was working at a radio station. There, we were forced to think about how we could tell a story in three minutes while staying relevant. So it wouldn't be about Robin Williams in Berlin, but maybe about Koma Saufen or abortion, all within three minutes.

I asked myself what I could do on Facebook or on Twitter to tell my story. My idea is now to provide a glimpse of one minute per day of my life to people without disabilities. My idea is to show these awkward moments, along with funny moments. Like when you enter a bus and everyone is looking at you. Or a nice moment when someone in a bus is helping you get the ramp, instead of the bus driver. The person is just doing it voluntarily, without being asked.

That's something I want to show to other people: what happens in my daily life.

I'm trying to continue writing the book on Facebook, but I tell more on Facebook with these entertaining, but also serious moments. I'm trying to make a mixture, without asking for pity. I'm just asking for empathy.

Talking about inclusion: do you think that the German government is lacking on bringing awareness to this topic?

RAÚL: Compared to other countries in Europe, Germany is lacking a lot.

We don't need to start a political discussion, but I believe that the most relevant topic on inclusion in Germany is that people without disabilities don't have enough contact with people with disabilities. While 10% of society has a disability, 10% of our friends don't have a disability. We have to ask ourselves why. And the reason is that there's an exclusion of people with disabilities in our educational and kindergarten systems. This is done systematically. We have Sonderschulen; we have Förderschulen. It is a big problem.

In Canada and also in Scandinavia, they closed all the Sonderschulen and Förderschulen and put the money into regular schools, just to mix people and create a bigger variety of people with and without disability in one classroom and in one school. Statistics show worldwide that people perform better in regular schools together with people with disabilities. They have better social skills.

On Förderschulen, not everyone is able to absolve the A-Level or the Abitur, although he has the capacity. The school doesn't provide that qualification. We have really a big educational problem in Germany.

Geraldine, as a part of Digitale Gesellschaft e.V., you're dealing with digital literacy and educational systems. Are you running into similar difficulties in reaching governmental offices to encourage that change in education?

GERALDINE: I thought you were going to ask a gender question now. Maybe I can answer both.

Everybody knows that we still have a lot of catching up to do when it comes to school systems and integrating media education. We need to integrate all the ethical aspects, in all the different subjects in school, from political classes to ethical classes, to religion. We need to address all the different implications about teaching people how to code in schools.

There's still a lot of catching up to do when it comes to the integration of digital aspects into our education system. The pressures is there, because of "Industrialisation 4.0" and the related German government's campaign to promote digitizing the German economy.

I thought you were going to ask me about a gender issue.

I'm deeply traumatized by the experience I had throughout the last week at CeBit. It was horrible in many, many ways. But experiences like that also make me realize how some of the Internet community bubbles I work in are so advanced in gender issues and trying to create a gender balance of speakers. If you look at the Chaos Communication Congress and re:publica, they're inclusive of all kinds of people with different life views and different ways to lead their lives.

> I was told I was chosen because I am a woman, and not because of the content that I brought to the table. <

That makes me very proud of my community, but it also makes me very angry with other communities. I was in so many drive-by situations, like drive-by conversations involving old men in suits talking about the hostess standing right next to them in a really derogatory way. I was the only woman on stage for four out of five days of the week, and I was told that one of the reasons I was chosen was because I am a woman, and not because of the content that I brought to the table.

The fact is that nobody thinks about it. It's not even an issue. It's nothing that people want to discuss. Every man that I talk about it with at Cebit immediately has 20 good reasons why introducing quota is a terrible idea. I'm regularly so shocked when I walk outside of my little bubble into any other media or IT bubble, finding how archaic things still are in this country.

You only can hope for some spillover effects, if they look at re:publica and other conferences.

GERALDINE: There needs to be great political change, and it's time to stop waiting for something to happen by itself. We need action that will lead to change, because it's not going to come about otherwise in the regulatory sense.

RAÚL: I totally agree. I have the same feeling.

What do you think about quota, dealing with either women or disability? A lot of women say no to quota for women, and then there are men saying we need a quota to regulate ourselves. How do you feel about that?

GERALDINE: I am definitely for quota, because there needs to be a regulatory framework for things to change.

Do you think the right people will then be nominated for a position, if we have a quota?

RAÚL: Let's consider the statistics. Women perform better at education than men. If you are really looking for skills, women should already be dominating us, but they aren't.

If you use a quota, some women might be in the wrong places, but there are a lot of men who are already in the wrong positions. We just have to start doing it and using it as a *Brückentechnologie*, for example.

GERALDINE: Yes, that's exactly the term I was thinking of, too.

RAÚL: We'll use it until it's not needed, but we need it now.

For example, when there's a 30% women quota at DAX companies, but there are only 300 jobs in Germany, that's nothing. That is half of all jobs at ImmobilienScout24.

I came across an interesting idea a couple of weeks ago. Somebody drew parallels between the black activism movement in the '60s and '70s in the U.S. and movements today like Arab Spring.

She pointed out the social coherence created by the fact that those activists organized themselves in the '60s and '70s without having all the tools we have today. They grew much stronger than we see today in similar movements.

This relates to resilience within groups, allowing a group to react better to new circumstances. It seems this resilience is missing today in movements like the Arab Spring. What's your take on that?

GERALDINE: I cannot give you a scientific answer with any opposing research. I can only give you a personal impression.

When I look at all my friends who live in Egypt and were involved in both attempts of revolution in the Arab Spring, I do not think they lack resilience at all. Also, with the activism we do here in Germany, the NGO that I'm involved in, Digitale Gesellschaft e.V., and also all other political NGOs, I think we're really resilient, too.

I don't think I would necessarily agree with that assumption, then. I think there are many factors making the world very complex. It's sometimes very difficult to keep up your engagement. If you're faced with brutal violence by the regime as a reaction to your activism, like in Egypt, of course it's hard to keep it up.

RAÚL: It is very hard to compare the black activism in the '60s with the Arab Spring, because people in the United States of America lived in the '60s as a democracy. The Arab countries don't. I would guess that it is or was easier to change or to fight for equal rights within a democratic system than totalitarian.

> Change is gradual,
but it also swings back
and forth. <

The conditions to provoke change within a democracy are easier than to provoke change in a terror regime, with a lot of torture and death sentences. I have the feeling that the conflict in the Arabic countries is more complex than the "racial discussions" in the U.S.

GERALDINE: There are so many ways that you could go into this. Another comparison you could make is why did the Occupy Movement not go further in evoking actual change in the U.S.? I think that was part of a long-term movement of change of perspectives within a system, where a few people hold the power so tightly.

Of course, the black power movement created fundamental change and equality was enforced in many areas. But in many, many other areas, the U.S. is still an extremely racist country, especially when it comes to systemic racism. This was shown in the Ferguson case and many others. Change is gradual and can have different ways of expression, but it also swings back and forth.

Are you experiencing a change in censorship?

When you do social campaigns on topics that are a little bit more sensitive, like ethical topics, is it changing how people discuss with each other?

Often, you don't have a person sitting in front of you responding, so it's easy to just throw something out there without thinking about the effects.

RAÚL: Before I was on YouNow, I thought that young people are more aware and sensitive regarding privacy, but now I'm not so sure.

A friend and his children are very careful using the Internet. They never type any email addresses, phone numbers, names, or even their genders. They're using the Internet more passively and anonymously.

On YouNow, I saw the girl that I mentioned earlier, showing her rabbit, her parents, and her whole house. It's so wrong on so many levels.

GERALDINE: This is another gradual issue. I have different impressions of this, too.

On the one hand, I would say that in a Post-Snowden world, the awareness for privacy and data security amongst activists has increased. I see more people using email encryption and anonymity tools. They're people who aren't necessarily Internet activists, who just decided to look into this because it's in the general field of their democratic interest. They want their privacy to be protected.

On the other hand, that reality that Raúl just described is definitely true as well. Of course, we've had many debates about this and the "*I have nothing to hide*" mentally. However, I reject the notion that society doesn't care. I think that is a great rhetoric that politics also use to wind out of the limelight, but I don't think that is true at all.

I have many conversations with people from all different traits of life, from taxi drivers, to people working in medical care, to my mom. I know that these are issues a lot of society thinks about. Maybe we aren't seeing those mass demonstrations, but it voices itself in different ways.

Let's close the session with a very general and maybe hopeful question.

What is the social change that you're most looking forward to in the next year? What is it that you're working on and that you want to see happen in society?

GERALDINE: On a positive note, we have seen an increased awareness for Internet freedom issues amongst different areas of politics. We see more people within different parties and within different areas of government become aware of these issues, understand these issues or at least try to understand. And that is a positive direction.

Personally, my fight in my area of developmental politics involves creating access in different areas of the world. More and more people are joining the Internet in our global society. I'm trying to create awareness for the importance of understanding the consequences of technical development in the sense of Internet freedom in development politics. That is a very, very tedious game.

As an example, a couple of years ago, I had people from a big development corporation come to me and say, "We've been active in Egypt for 20 years doing educational development, but this whole Facebook thing skipped us. Can you explain what happened?"

Then you see the danger. People from the government can say, "We're going to support the projects for all activists who are active on Facebook."

> I'm hopeful that we haven't lost the fight to protect our free Internet, not just in Germany, but globally. <

Then you try to say, "Well, wait a minute. That's just one aspect of it, but you want to understand issues about privacy and about data protection if you're going to do that."

It's very, very tedious. Slowly, I see things moving. I'm hopeful that we haven't lost the fight to protect our free Internet, not just in Germany but globally.

RAÚL: I'm looking forward to using the Internet to reclaim some definitions on the topic of people with disabilities and inclusion.

In the last years, I realized that mostly non-disabled people are talking about people with disabilities on the Internet. That's a big problem. It's like men talking about women and saying what is best for women, without being a woman.

I want to use the Internet to give people with disabilities the power to raise their voice. We want to help them toward self-empowerment, so they can fight for their own rights.

Maybe we can make some connections and reclaim the definition of what is inclusion. We don't want to give it away to teachers who fear having too much work or parents of non-disabled peoples, who only argue based on their fear of something they have never experienced themselves, based on no real arguments but on emotions.

I hope we find a solution. I hope we don't argue based on emotions and wrong preconceptions.

Today, we heard it's tedious; it's a lot of work. But everything we do will further that change.

Today we had a glimpse into your world. Hopefully, some of us found inspiration to look for causes or start our own.

> I hope we don't
argue based on emotions
and wrong preconceptions. <

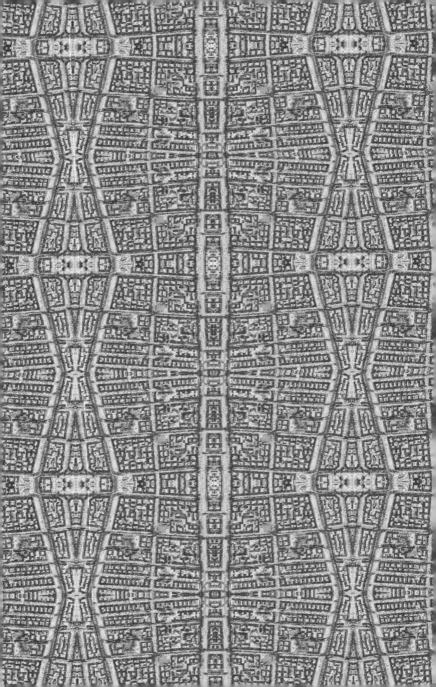

4
New Work Order: Who is in Charge?

Cristina Riesen
Monika Frech

PANEL DISCUSSION, BERLIN,
JUNE 10, 2015

The concept of work and how we
define it has changed dramatically in
the last two decades. In the wake of the
industrial revolution, employees
committed their whole careers and lives
to one company. Now, Google and
Amazon report that the average tenure
of an employee is just 13 months, not
33 years. Modern knowledge workers
are choosing flexibility over security and
stability. Technologies allow employees
to educate themselves and build knowledge. Thomas Malone of the MIT Sloan
School of Management refers to this
as "a new era of hyper specialization,"
which allows freelancers to build their
own "Specialist Brand."

Sourcing talent globally is common
practice. To handle on-demand work
tasks, we seek out tools and smart
technologies that create ambient knowledge exchanges between people and
across teams. This allows us to work

more effectively and efficiently. But how do companies maintain culture and foster effective collaboration when the designer is in Zurich, the engineers are in Austin and the launch is in Beijing? How do we manoeuvre through the myriad of apps, programs and digital tools to create the most productive environment, stay connected with peers and manage our knowledge and expertise?

Cristina Riesen, the General Manager of European Evernote, and Monika Frech, one of the 30 co-founders of Dark Horse Innovation, address these and other questions in this panel discussion. Both have dedicated their work to improve the professional lives of knowledge workers.

PANELLISTS

CRISTINA RIESEN is General Manager EMEA at Evernote. In this position, she champions the modern workspace, working with companies as they move away from the traditional understanding of sharing files, collaborating and presenting. In 1996, she started her career as a radio editor at Pro FM, the first private Romanian radio station. After graduating summa cum laude from Transylvania University in Brasov (BA English and French), she moved to Switzerland and continued her studies in public relations and corporate communications at the Swiss Public Relations Institute. She is a certified Reputation Institute Professional and graduated in November 2011 as an Executive Master of Science in Communications Management of the Università della Svizzera Italiana, one of the leading programs of its kind in the world. Before joining Evernote, Cristina was involved in several online community management, social media strategies and business development projects for the Volvo Group, Marvin Watches and Engagor.

MONIKA FRECH is co-founder and partner at Dark Horse. She is part of the marketing team, spending half of her professional life speaking at various conferences and the other involved in an NGO for Indian children. As a co-author of *Thank God, It's Monday*, she explores new work paradigms and ways to improve company innovation culture. Dark Horse uses Design Thinking, Service Design and Culture Hacking to enable organizations to utilize the market and work potentials of the digital age. The agency was co-founded by 30 Generation Y professionals from 25 disciplines. They work together with flexibility, without any hierarchies, but with a shared mission: To innovate and do things differently.

Monika, today we were with Cristina at the office of Dark Horse[109]. We discussed the unique corporate concept of Dark Horse: 30 people starting something, no boss, and no structures.

Or are there structures? Please, tell us. What was your motivation to create this, and how did it develop in recent years?

> MONIKA: We started out as a group of students. All 30 of us studied together at the HPI D-School[110] of design thinking for one year. And after that one year of studying, we became good friends. We also start liking the way of working we'd been introduced to there. After that year, we thought, that couldn't be it: let's do something together.
>
> But we didn't set out and plan to become a company. It was rather the other way around. We realized we have to earn money with the stuff in order to stay together. Otherwise, people would go off to other jobs, or do different things.
>
> We accidentally became a company, and today we call ourselves Agency for Innovation Development, only for lack of a better word. If you know a better one, please let us know, since we're still not really satisfied with the word.
>
> We've been doing this for six years now. We basically help companies innovate products, services, and their company culture. We mostly learned how to do that from our own failures during the last five years.

So what is this new, modern, working space for you? Cristina, you are providing the tools for them.

109   *Dark Horse is a team of 30, who work without hierarchy, constantly innovating the best ways to work, continuing their education, and altering creative spaces for best workflow.*

110   *The HPI D-School in Potsdam is the European center for a growing Design Thinking community.*

CRISTINA: As just mentioned, today we were visiting Monika and Dark Horse, and it just hit me: This is the future of work. This is the modern workspace, because it is all about sharing knowledge, coming together to build something truly great, that has an impact.

If you look at knowledge workers today, they are not necessarily driven by money and benefits. It's more about ideas and having an impact. Coming together, working with smart people, in a flexible environment that adapts to their needs. Together, they are making a difference.

> This is the future of work. This is the modern workspace. <

When we designed Evernote[111], this is exactly what we had in mind. We actually built Evernote for ourselves. We started seven years ago. We were a bunch of friends, overwhelmed with this crazy amount of information. We thought there must be a smarter way to deal with this, and to work, and to just manage all the knowledge around us. We thought, if we came up with a solution, there was a high chance that millions of people out there would like it too.

Today, we have 100 million registered users worldwide. Evernote has been growing mostly organically, via word of mouth. People found it very useful.

Through working with Evernote, you probably saw the value shift that goes along with this paradigm shift in knowledge work.

We have more and more knowledge workers; what can you say about that?

111   Evernote is a software and web application that helps you collect, organize and find notes, documents, and photos in a variety of formats.

CRISTINA: A couple of years ago we noticed that two-thirds of our users, approximately, were using Evernote in a work context. By doing some research, we've noticed that, despite their bosses and managers prohibiting them from using modern technology, people were using it anyway, because they were able to do a better work. At the end of the day, it really involves the question of how smart companies will be successful in the future.

First of all, companies need to trust their people. When you hire somebody, you really trust that they are absolutely great at what they are doing. You trust them, empower them, give them the right technology, and then get out of their way. Let them do their thing. It's not like you have to micro-manage them and force them to sit at their desk from 9 to 5.

This has to do with the way we humans live and work. It's not like our brains will necessarily think business from 9 to 5, and afterwards you'll switch your brain and you'll start thinking about hobbies and families and anything else. It's really a natural, on-going flow.

With flexibility, you need trust. Trust becomes a central value in a modern working surrounding.

CRISTINA: Yes, it's fundamental.

How does this work with the value system at Dark Horse, especially regarding your book *Thank God, It's Monday*?

MONIKA: Trust is one of our fundamental working principles, but it's not a given. We were really lucky to start out as a group of friends, so we already had an established trust level, but it's hard work to keep that trust level.

We actually put a lot of work and effort into maintaining the trust to each other. Of course, with 30 people, no hierarchy, no boss, no one making the decisions; of course we fight. But at the end of the day, we all know that we fight for a good reason, and that we don't fight each other. We can still keep the trust in each other as people, as knowledge workers.

How do you manage the balance between intrinsic and extrinsic motivations?

Obviously, everybody like to be paid, everybody likes to be valued, and these are things that you experience in your own structure. However, you are working with big corporations, and helping them to innovate.

What is the difference between the dynamics in Dark Horse and the dynamics in corporations you work with?

> MONIKA: A lot of our clients still rely on the extrinsic motivation, such as all the classics like bonuses, and hierarchy levels, and then growing bonuses, and so on.

> For us, because we started out in that start-up way, at the beginning, we had no money at all. Here were 30 very young people, who really liked to work together and do great work, without really knowing what the work would be.

> In the beginning, it was all intrinsically motivated. But of course, we also have to live, and of course we also want to live a good life. Yes, we also need the money.

> Through a lot of iterations and a lot of experiences and ups and downs, I think we've now found a way to balance the two, for now. That's actually one of our struggles, and it's not set in stone.

> We keep working on it. How do we compensate for time, and how do we compensate for tasks that nobody really likes? Things like that.

How does the need for structure and rules go along with values like flexibility and meaning? Cristina, you mentioned that meaning has a higher value in the new working surroundings.

CRISTINA: At Evernote, we are based in Zurich, running the EMEA[112] operations. It's around 20 people. You can clearly see that the turnover is very low. There are people who have been there from almost the beginning, and they are still there, because they found meaning in what they are doing. Every day, they have a huge impact on millions of people's lives. This is a very strong motivator in our case.

When I did my research, I stumbled upon a quote of Peter Drucker[113], who had already foreseen that knowledge workers would rise in the future working society. He said, decades ago, "The most valuable asset of 21st century institutions, whether business or non-business, will be its knowledge workers and their productivity."

How can companies unleash this internal and external potential of knowledge, this innovative potential? Monika, at Dark Horse, you work with companies and consult them on this topic. How do they deal with it?

MONIKA: These companies all read Drucker and this management literature. Until today, it was all very much theory. But now we see that a lot of big corporations try to walk the talk. Of course, it's very hard for them to break up 40, or even more, years of structures and the way they worked.

> "The most valuable asset
of 21st century institutions will
be its knowledge workers
and their productivity."
Peter Drucker <

112 EMEA: Europe, the Middle East and Africa

113 Peter Drucker was an Austrian-born American management consultant, educator, and author, whose writings contributed to the philosophical and practical foundations of the modern business corporation.

To us, it comes down to principles instead of fixed rules. The principles are mainly collaboration and iteration: you should experiment, not have a set strategy or rule, and then go for it, no matter if the circumstances change. We focus on user centeredness. In a company, one of the sets of users are the employees. You should really try to see what works for your employees at any given moment. That's something very far from corporate reality.

CRISTINA: A couple of years ago, we launched Evernote Business[114]. One of the ideas behind Evernote Business was to surface all this hidden information inside organizations. Even if you are a very small team, it is not humanly possible for you to know what your colleagues know. You're spending a lot of time just starting a project, doing research, only to find out that somebody sitting two desks away has already been working on that. This shouldn't be the case.

In order for you to be productive, smart, and to make good decisions, you should have great tools that allow you to have easy access to whatever information is floating around in your company that can be valuable for your goals.

Also, we were frustrated with business software in general. We thought: why should you be punished with lousy technology at work, because you spend most of your life working? It shouldn't be like this. Actually, you should have a great time when you work, and you should have the same user experience. You should not feel like technology is controlling you or you're forced to use it. It should be doing a service.

---

114   With Evernote Business, you're offered the ability to save team projects in one place for ease of working with others, share content and give feedback, collect pictures, files, and Internet searches, and synchronize all computers and smartphones.

Before, you mentioned the term AI[115], interpreted in a new way as augmented intelligence, which gives the user super powers. You don't say it's a tool, but rather something that works in the background. I'd love to hear more about how we can use it in a positive way.

At the same time, in Germany, we talk a lot about data protection and surveillance. You were born in Romania, which is, like the Soviet Union, a very repressive state. Being watched and having all the data accessible for everybody at all times has its ups and its downs. How do you react to that?

CRISTINA: Let's start with artificial intelligence vs. augmented intelligence.

When we started building Evernote, it was clear that the way we perceive technology is again like having super powers. Technology is not there to take you over, or to think for you. It is there to be a natural extension of your needs. Technology should be there to anticipate what you need in whatever context.

With Evernote, which uses augmented intelligence, you're basically stepping into a meeting. If Evernote is connected to your calendar, it will suggest a name for the meeting. It will add whatever relevant information, without you actually having to start searching and freaking out. Everything is there: the article you've clipped, your business card, and so on.

It was all about allowing you to focus on what is really important, which is creative thinking, decision-making, and coming up with new ideas.

There was a lot of discussion around artificial intelligence for a long time. Remember when IBM's supercomputer

115   *Artificial Intelligence is intelligence exhibited by machines. In computer science, the field of AI research defines itself as the study of "intelligent agents." It's any device that perceives its environment and takes actions that maize its chance of success at some goal.*

Deep Blue played chess against the world chess champion Garry Kasparov[116]? Garry Kasparov had an epiphany. He realized that actually, the future of humanity is not necessarily man vs. machine. It's actually man and machine, working together, to make man smarter.

> The future of humanity is not necessarily man vs. machine. It's man and machine working together to make man smarter. <

Back to the other topics of data security and protection. From the very beginning at Evernote, we made a very clear strategic decision, when looking at how we want to build our business and how we want to make money. For us, it was a long-term vision to give everybody a second brain, to make the world smarter. We knew we couldn't achieve this if we started selling people's data. From the very beginning, it was very clear that we could never monetize on people's data, although it's very appealing. In the short term, you can make a lot of money.

For example, if I look at what you're sending into Evernote, then I can sell you a certain product. But we decided not to do so. This is why we say we are the 100-year-start-up, because we believe that we can make a difference on the long-term. If we want to gain people's trust, we cannot play tricks, and we cannot do whatever we want with their data. It's one of our top priorities.

When you put information into this, you have the opportunity to work in virtual teams. But if somebody leaves the team, the knowledge stays with the company.

116   Deep Blue versus Garry Kasparov was a pair of six-game chess matches between world chess champion Garry Kasparov and an IBM supercomputer called Deep Blue. The 1997 match was the first defeat of a reigning world chess champion by a computer under tournament conditions.

How do you create loyalty and commitment within this atmosphere of, "We already have your knowledge, and we can build on it anyway," without you?

> CRISTINA: Going back to when we launched Evernote Business, there was a third important idea behind it, which was again in terms of the user experience.
>
> We asked: Why should you go outside of an application to access your personal notes, when you can have everything, personal or business, in one place? We wanted to have this one interface, so you wouldn't have to change anything.
>
> Then, the moment you left a company, all your business notes stayed with that company. This is not done to control you, because you can easily copy and paste everything into your personal account if you want to. It was not done for that. It was because, as a company, especially with people coming and going more and more, you want to make sure that the knowledge is preserved. You want to make sure whoever joins the company has easy access to whatever projects have been done before. This is helping a lot of companies create this new modern workspace.
>
> Loyalty has to do with believing in a mission. You can be loyal to a company because you truly believe that you can make a difference, and you embrace the vision and the goals.

If we define a company as the sum of committed workers and the sum of the information a company creates, in a sharing society where knowledge and talents are sourced globally, what is the definition of a company?

How is business going to change? How is the definition of a corporation going to change?

> MONIKA: For us, early on, we decided that loyalty isn't something that you can enforce. That's why we created

an open system. Since we didn't find any system in the corporate world, we used an analogy of monarchies. At Dark Horse, once a year, on our so-called commitment day, you can choose whether you want to be a monk or a nun for the upcoming year, or go on a pilgrimage.

Have you been on a pilgrimage?

> MONIKA: I am right now, sort of. As a monk, you commit most of your time, or some of your time, to working at Dark Horse, and you get compensation for that.

> You can also say no, that in the coming year, you want to build a house, you want to start a family, or you want to work in another job. Then you are on what we call a pilgrimage, but you're not thrown out of the monarchy, or out of the company. In the next year, you can decide whether you want to come back and work as a monk or a nun again, or continue your pilgrimage.

> You're always still part of Dark Horse. You can work in a different job. You're welcome to take your bag of experiences that you had on the pilgrimage and come back.

> That's another way of creating loyalty, not by enforcing it, but by keeping the doors open and inviting people to make different experiences. For us, we discovered it's really important for the way of working, because if we want to stay innovative, we cannot just close the doors and be with ourselves and not have different experiences.

So you could stay with Dark Horse for 30 years, changing companies in parallel?

> MONIKA: Yes, maybe. We don't know yet, but yes, hopefully.

I was wrong in the beginning with my outlook that people would stay with a company for 13 months on average. We might have new structures, where we stay with a corporation for 30 years, like with Dark Horse?

MONIKA: Yes, 30 years with Dark Horse, and 30 different jobs, maybe.

Corporations would become like hubs, maybe.

You work with big corporations and projects. With Audi[117], you are trying to help them innovate and get the teams out of their regular thinking patterns. Are they open to these kinds of structures?

MONIKA: We find a lot of openness on the C-level as well as on the team level, where the actual work gets done. The middle management is very reluctant, because they fear losing ground. It's actually hard to convince them.

In our experience what works best is having support from the top management, and then just making small steps with individual teams. We then show that it works. We show that it's not just a fancy work tool. People like it because it actually really works. It actually fits the 21st century. Once people realize that, then usually it's easy to convince people who were reluctant before.

Cristina, you run a very international team, with many virtual workers around the world. Your designers and developers are in different places. How does that play out?

CRISTINA: Building a decentralized and fluid company culture, and keeping people together, doesn't just happen automatically. You have to work really hard to make it happen. It's all about team building. It's all about making sure that you're encouraging differences of opinion.

Internally, we are doing this a lot, because we are a bunch of very different people, different nationalities, and different

117   Audi AG is a German automobile manufacturer belonging to the Volkswagen corporation.

backgrounds. It's only by the nature of things that sometimes you'll have clashes. People will feel that, saying, "My idea is better, and I want to push it through."

You have to help them. They have to take a step back and realize that it's great to have differences of opinion. If you're falling into this groupthink environment, where everybody loves each other, and they agree on everything, not much excellence will come out of it, not long-term.

You must express those differences and focus your energy into finding solutions to these challenges. Don't direct that energy against each other. Use that energy to find new creative ways to solve problems. It's an intensive process. I think this is the new kind of leadership, especially for the mid-level.

MONIKA: In our experience, true empowerment does not come from making decisions, but also from deciding on how to make decisions. Deciding on the sort of governance that a team or a company has. That's the highest level.

Once teams are abled to decide upon their own governance, then you have true empowerment. People are really passionate about their work and really involved.

What do you think about collaboration vs. competition, in your working routine as well as on a feature level?

Today, people promote collaboration as a better way; but competition has its values regarding innovation management. What do you think about that?

CRISTINA: At Evernote, our philosophy of life is around collaboration. We do not believe in competition. This is not meant in an arrogant way. We believe that life is not a sportive game, where in order for us to win, somebody else has to lose. It's more like playing in an orchestra, where you have different people playing the same instruments, but together you're doing something amazing.

This has played well for us. People ask about our competitors, and then we say, look at our partners; they are Apple, Google, Microsoft, and Moleskine.

This was only possible because we thought: why should we even waste time looking at the competition? Why not focus our energy into talking to those companies and together try to come up with something truly great.

As a team, we also have this approach. We believe that again, collaboration can be super beneficial. Competition is a bit tricky, because you end up in politics, especially as an organization, which involves a lot of complications. Then, it's not a healthy environment.

I think having this kind of flat structure, empowering people, and really fine-tuning every day, is key for us. We ask: how can we do better, and how can we make sure that people really feel empowered and responsible for what they are doing?

MONIKA: I can second that. It's hard for us to compete against each other within our work, because we are from so many different disciplines. My background is in writing, so I cannot compete with my colleague who is an engineer. It just doesn't make sense.

We realize that we sometimes lack that element of competition, and so we introduce it in a lot of playful elements. This comes back to establishing the culture of trust. We have an internal games league, and we have the tabletop football, that sort of stuff. We're mindful to have that playful competition, because for us, it didn't work out without that element.

At the same time, the more freedom you give people, the more you need to check if they're reaching their goals. You have project deadlines, you have scope, you have budget, you have deliveries, MVPs[118]; so you need to track what people are doing.

The more freedom you give them, do you feel you also have to find ways to track what they're doing? A friend once used the term co-opetition, somewhere between collaboration and competition. How do you manage that?

> MONIKA: Well, we don't manage that. As you said before, we really don't have a boss. There's no management level, nobody who controls what everybody else does. We do it ourselves.
>
> When we start on a project, the project team is responsible for everything. For budget, for planning, for everything. if you're not on time, if you're not on budget, it comes back to you as a team. Of course, it also comes back to the whole company. Mistakes have been made. Many of them.
>
> Coming back to sharing knowledge, we realize that if we don't share what doesn't work, then we don't actually have trust. We have an internal fail award that we give out regularly. You can only nominate yourself, so it's not a blame game. You have to step up and say okay, guys, friends, listen, I did this.
>
> By sharing what didn't work, we can learn from each other and hopefully not make the same mistake again. It's sort of a circle. You have to share your failures to get the culture, but you need the culture to be able to share the failures.

Evernote probably has a very decentralized structure, with the developers separate from the marketing guys. Do you track their results at all?

118 Minimum viable product is a product with just enough features to satisfy early customers to try to provide feedback for future development.

CRISTINA: We do. Basically, it starts with how and who you're hiring. When you hire the right people, people who are self-starters, you know that you own your topic from the beginning. This new person is here because he's great and he's going to figure out how to contribute to us reaching our goals.

I think it is super important and critical to define goals as a company. The way to track it afterwards is dependent on the department, in our case.

I can give an example with my team in Zurich. When we have team meetings, we come together, we see where we are, we check where we are, what's needed, what we can do better, and then everybody leaves to do their thing. We come together in another week, and we kind of see again where we are, and we fine tune, from one week to the other.

It's not like I'm sitting there waiting for everybody to report to me what they've been doing, and how it's going. That doesn't work.

It's a team coming together. Everybody owns his or her topic. We interact to figure out independently what could work. If you're working in sales or marketing, it doesn't matter if we're discussing user experience. We want everybody's opinion. It's about really being part of the same boat.

I think in our case, we've taken a lot of steps into making sure that we ban bureaucracy in general. We have these very explicit rules against, for example, stupid meetings. What is a stupid meeting? A stupid meeting is the one where you come in, you don't know what's happening, you stay there, you write three emails, you send a message on Whatsapp, and then you leave the meeting. You have no idea what's going to happen afterwards.

>We have very explicit rules against stupid meetings.<

Whenever you're booking time on somebody's calendar, keep in mind that you're taking one hour away of that person's life. It's a huge responsibility, and don't mess with it. Think very well, is that meeting really needed? Can you solve it otherwise? When you do call for a meeting, be prepared. Send an agenda, say clearly what is expected, and then work it through.

You can become easily drawn away from these tools as well. Personally, I use Trello, Slack, GoogleDrive, Skype, Evernote, and sometimes I feel crazy.

What would you recommend to a knowledge worker like me? How can you make my life easier?

CRISTINA: It's a personal choice. I think it's a curse and a blessing at the same time. With all these varieties of tools, you can look at them as a blessing, because you can actually select and customize. You can create your own Swiss knife solution. Depending on what you want to work on, you'll have different tools.

I know that it's also a curse, because you get lost with so many tools appearing every day. At Evernote, we also use a lot of these tools, because Evernote doesn't solve it all. But it comes down to making sure that you keep up. I think it's important to keep up with the technological development, but really being very mindful about technology not taking over you.

Today, if you're looking for a message, you ask yourself: was it on Slack? Was it on Gmail? Was it on Facebook? It's up to you at the end of the day to be really disciplined. Personally, I switched to Slack, so whoever wants to follow me or reach me, the best way to reach me is on Slack. Period.

No emails anymore!

MONIKA: Sometimes, the best tool is to grab the phone and call somebody, or meet in person. A lot of our clients are on the same floor, but they always send emails to each other, and it goes back and forth for days and days. You could just get up from your desk, walk to your colleague, and get it all sorted out in maybe five minutes. It's the culture that you have around communication that defines how you use the tool. You can't let the tools use you.

At the same time, organizing these tools costs time. Are we moving from exploitation to self-exploitation? Meaning will we end up working more, because we're freelancers, knowledge workers, and we have these goals? We sit at home, we sit in the park, but in the end, we still use the laptop, the tablet, or a smartphone to continuously communicate.

In Germany, work-life balance is an issue. How do you enable people to actually have a better life-work balance? Is that something you aspire to at all?

CRISTINA: We always speak about work-life integration rather than work-life balance, because why should there be this contrast?

At the end of the day, technology is empowering us in ways unseen before. It's the very first time in history where we can get back our lives. This comes with great responsibility, and this is where we need to step back and find a new way to live and work, by taking advantage of technology. Not letting technology take and run away with our lives.

It starts with simple things like deactivating push notifications. Why should you know whenever somebody is sending something? It starts with simple things like thinking about what you want to achieve, and what is the best way to do it. There are millions of ways in which you can use technology to work for you, to have it as your private assistant. This means a new way of looking at life. We need role models, people like you here, to be more vocal about the best way to work. How can we actually use technology to our advantage?

Just to give an example, when I started working as a PR assistant, in Bern in Switzerland, that was some years ago. I had a very small kid back then. Now, she's almost 15, so it's much different. Back then, I was working in this agency, I had to be in the office at 9, and my boss would not accept any five-minute delay. You had to show up at 9 o'clock. If not, you were promptly fired or something. I was rushing through traffic with a kid, because I had to go to the kindergarten. Sometimes, it was just almost impossible. I was stressed out completely when I arrived because I was trying to figure out how to get to work in time.

At work back then, as a PR assistant, the first thing you would do was you would take a pair of scissors and start cutting out articles from newspapers, depending on the client you had. Then you would clip them into this huge folder. Your boss would ask if you could please find an article with the CEO from 3 years ago. You would literally go and flip through all the stuff.

Today, if I was a PR assistant, my life would be much easier, and I could actually focus on more meaningful things. I think this is where technology is helping us, because you can spend more time focusing on improving your skills, learning and doing something that has to do with creative thinking, rather than just repetitive tasks.

MONIKA: I think a lot of it comes back to user centeredness, because often, the corporations take the one-size-fits-all-approach. All emailing is switched off at 5 p.m. For me, for example, I have a baby at home, and I send emails out to my colleagues and clients at 11 p.m. They all know that they don't have to reply at 11:05 p.m.

I won't reply at 9 a.m., because that's when I'm busy. They know that I do reply, but they have to wait for it. If the emails were switched off at 5 p.m., it would be a disaster for me, since I just couldn't work anymore.

The solutions chosen by big corporations for all of their employees is sometimes not so great for all. These one-size-fits-all solutions are often a big mistake that I've found often in Germany.

Is there also the risk of a digital divide between workers, because one person knows how to use these new tools, or knows how to deal with communication technology, and the others don't? How can we overcome this gap?

CRISTINA: There's certainly this issue, but I think, depending on the industry you're in, if you're looking at adoption of technology, it's kind of happening organically, if a tool is becoming popular, you have the power of the crowd. A lot of forums and discussions and people are vocal about how that specific tool is helping their lives.

With some tools, like Slack in our case, everybody suddenly from one day to the other, everybody was using Slack. We had to give it a try and see what it was all about.

With tools like Evernote, it's a bit trickier, because it's rather complex. You wouldn't understand what you could actually do with it in five minutes. It requires you taking time, reading some articles, and experiencing it a couple of times, so it is an investment. There is a risk that you're missing out on tools, just because it's not so easy to adopt them. This is where I think, we, as companies, and we as Evernote, need to do a better job. We need to make that user experience, especially at first launch, better, because we haven't been that great.

MONIKA: At the same time, the tools won't solve your problem per se. We saw that with a lot of our clients, who thought that once they implemented a new tool, everything would be done. But then they say that the tool doesn't work. But you actually have to enable people to use whatever new tool you provide or create.

Yes, you have to establish a culture around it.

MONIKA: Yes, and also, if you are just implementing the newest stuff on the market, you'll always be behind, because you need to have employees or students who are prepared for what will be next, and not what's there now. That is a big mistake that a lot of companies and the education system in general are making.

When you're hired today, you at least need to be able to manage office products. But now, you have Google apps, then you have Evernote, you have all these different tools. I was talking today to an agency that is promoting Google apps for corporations. In the end, they have it all installed, but only a few people actually know how to use it. It's like driving in a Ferrari, in the first gear.

How do you deal with that when recruiting, especially as people don't stay with companies that long? You invest in these people, you teach them, and then they move on. Then, you have the next generation that you need to teach, and as we said, the school system also not preparing you for this.

CRISTINA: In our case, I guess we are lucky, because we developed a way to retain knowledge. So this is independent of people are coming and leaving. We are seeing this as the new normal. It's not shocking, if people come and stay for only a couple of months, and then move on. What we want to make sure is that first of all, the time that's spent with Evernote is a great experience for them and that they are learning. If they are happy and learning, in those few months, they will contribute significantly. We try to create this infrastructure that allows for this very rapidly shifting workforce.

In Zurich, for example, and also in other cities around the world, we have people who have been with Evernote for quite some time now. Sometimes we see people coming and going, and we are prepared to deal with that. The goal is not necessarily loyalty, at the end of the day. It's: how are you creating an environment where you can come up with amazing ideas and execute them, having different players?

I think it has to do with the future of work in general. It's no longer a question of companies; it's more like structures of teams coming together. It's the new freelancing economy.

With Dark Horse, this is the perfect example that it's actually working. You have different people, different backgrounds, coming together in a very unorganized way.

At Dark Horse, you are really trying to understand and focus on Generation Y[119]. They are defining the work culture of today or tomorrow. They're also described as the Peter Pan Generation, jumping around, the Selfie Generation, self-centred, looking for meaning and coming and going.

How does Generation Y assess the work surroundings? What's the difference to Generation X?

MONIKA: That's the Peter Pan Generation; they don't want to grow up. If growing up means staying at one company for 30 years, and looking forward to retirement, and the weekend, and not working, then yes. We don't want to grow up. That's right.

119   *Millennials, also known as Generation Y, are the demographic cohort following Generation X. Demographers and researchers typically use the early 1980s as the starting birth years and the mid-1990s to early 2000s as ending birth years. Their upbringing was marked by an increase in a liberal approach to politics and economics.*

> If growing up means staying
at one company for 30 years,
and looking forward to retirement,
and the weekend, and not working,
then yes. We don't want
to grow up. <

Generation Y has been defined as the Selfie Generation, and always looking for a new opportunity. It's all about flexibility. But we think you can create an environment where everybody can thrive, and can actually live to that kind of selfie-ness potential.

We see a lot of collaboration. More than there was before. I would say that's the main difference to the Generation X[120], which has been described as just "me, me, me, and money, money, money." That's something that's supposedly not so important to generation Y anymore. I'm not so sure about that. We talked about the sharing economy, but somebody still needs to own all the stuff that is shared. The digitalization enables us to live the lifestyle of the baby boomers and of Generation X, without having to own all that. The lifestyle is still pretty attractive to a lot of us.

What we see with our clients, or what we actually know from five years of experience working in changing organizations, is that Generation Y is actually not very much of an artefact. People of all ages and of all backgrounds want to enjoy their work. They just have been taught for 30 years that it's not possible. They've been taught that fun is on the weekend, and fun is when you're retired.

120 Generation X, or Gen X, is the demographic cohort following the baby boomers. The birth years typically range from early-to-mid 1960s and ending in the late '70s. They were children during a time of shifting societal values, with reduced adult supervision as a result of increasing divorce rates and increased maternal participation in the workforce.

It's just that Generation Y doesn't take it anymore. We don't want to work in an environment that doesn't fit us, that doesn't fit our needs, that's not fun. I think that's the difference. We are in a position to actually ask for what we want. The need itself, that's not a generational thing.

You're also in a very privileged position. You were in a group of people, and you decided to create a brand together.

Freelancers today, and a lot of people who use Evernote and other collaborative tools, have to build their own brands to survive in this economy. They're in a hire and fire war, and they must find new projects. They must be strong, not only in their knowledge, but also in promoting that knowledge.

The "me, me, me" generation does have its merits, because you need to promote yourself, and they can. How do you deal with all these self-promoters? Once you hire them, they're still in that same mode? You have collaboration between a lot of people who constantly need to promote themselves as individuals and brands.

MONIKA: Give them great work.

Then they start to collaborate?

MONIKA: I think so, yes. They will promote themselves as a result.

CRISTINA: Give them great work. Not only that; but allow them to do their side projects, and salute that. Encourage them to do something else. If they want to write an article on Medium and do something else at the same time, let them—that is great.

When YouTube started, we saw a lot of people with a lot of opinions. These videos did have some strength and some curiosity. When you look at YouNow today, which is a new live stream channel, you realize it's all about self-promotion and a lack of content. We're looking at the next generation of only promoting. They don't have expertise. They're just promoting themselves as a brand.

> MONIKA: Somebody must be interested in watching all those self-promoters. There must be some sort of value in that self-promotion.
>
> Maybe they are giving some meaningful content. For example, there are those videos that teach you how to do your makeup, or how to sew your own clothes. That's obviously of some sort of value. But it still seems very self-promoting to just put on your makeup in front of the camera all the time, yes.

Let's switch to digital literacy.

Very early on, we need to learn how to deal with these tools. Have you both in some ways collaborated with schools and institutions, younger generations, and have any experience in how they perceive all these tools; as a threat or as an opportunity?

> CRISTINA: Evernote is very much used by students and universities around the world. I wish I had Evernote when I was in school. It allows you to take the knowledge with you as you grow.
>
> When I was at university, we still wrote everything on paper. When I moved out, my parents decided that's a bunch of crap, let's throw it out. Then everything was gone from one day to the other, nothing. Even if I wanted to go back—I had Old English, for God's sake, as a topic at university—I don't remember anything, like literally nothing. It's completely gone.
>
> Today, with all these tools, you can grow, and you can take the knowledge with you. This is extremely powerful. It's also very powerful in terms of teachers interacting with you on a more customized level, and also with the parents. There's a lot to improve on the school and parent communication side, as well.

From what we are seeing, more and more schools are working with technology, even allowing kids to take their own devices and access whatever when they are studying a specific topic, just go to a specific resource.

MONIKA: I think it's really important to teach kids and also adults not to use specific technology, but to live in the technical or digital society and to understand what digitalization does. It alters the way that we collaborate and the way that we work and the way that we learn.

That's different than having a tool and understanding how to use it. It means understanding why that tool exists and how people use it, and being able to create the next tool. Yes, you need some sort of technology for that.

From what I know, in German schools, technology doesn't very much exist at all. Yes, that's probably good if that changes. But you also need to teach kids how to be creative and how to be curious and how to discover things that they're really interested in. Maybe they'll create the next Google or Evernote.

CRISTINA: This is a great point. When you look at education, there's so much to be done, because we are basically preparing kids for a future that will not really be there. We are teaching them based on an old model, and it's not up to speed with what is happening today. We should go back and encourage creativity. I think creativity and everything that has to do with artistic performance—everything that has been seen as "not worthy" of school, in general, should be part of primary school.

If you think about technology and humans in the future, what is the biggest differentiator? It's really creative thinking.

This is what technology is not necessarily able to do. It's not able to come up with creative solutions. It will not have the dexterity humans have. You cannot have a robot work through a room and bring you a glass of wine without tripping. This makes us really special and powerful.

Again, we are still victims of this legacy of the post-industrial era, where education had to prepare this work-force for coming under one roof, and being controlled, and giving the output needed. This is no longer the case. We should stop killing creativity in schools.

It's not so much a question of empowering kids with technology. It's more a question of re-thinking education and school. It's about teaching kids that it's no longer a question of just going to the right school and then getting a job and then you'll be happily employed until you retire, because this is no longer the case.

Today, the difference is the variety of skills you learn and things you experience. It's also in coming together and having the possibility to have an opinion, and to express it, and to interact with others in a smart way.

Monika, you are an advocate for design thinking. When you say design thinking should be taught in primary school, I think about Picasso. He said that everyone is born as an artist. Is design thinking re-awakening this inner child?

MONIKA: It can be. Design thinking is a way of working together and creating products and services that really help people, or that are really desired by people. If it's understood in the right way, it can open up people's potential.

Very often, it's misunderstood: just another set of tools, just a new method. They try it out once, and, oh, it doesn't really work, surprise, surprise. The way we understand it is as a set of values that incorporates the design mind-set, and uses that to work on projects. Then it can actually work.

Cristina, how do you organize the internal body of knowledge at Evernote? How do you make lessons explicit and shared between your workers?

> CRISTINA: We were lucky enough to develop the tool ourselves. Sometimes, it's complex. Just think about tags, to give an example. If everybody starts coming up with tags, then you have trillions of tags floating around, and you cannot find anything. It needs some discipline.
>
> Sharing knowledge has to do with finding some structure, agreeing up front on how we want to structure our notebooks at Evernote. Everybody has a tendency to add everything everywhere.
>
> As a company, if you really want to be efficient, you have to have a structure. You must spend a lot of time from the very beginning, agreeing and making sure that everybody really is on the same page. How do we want to create notebooks, and who has access to what? The guiding principle is, for us: less is more.

How much do you use your own tool, and what happens before an expert meeting to share all your practical knowledge? Everything documented can be shared?

> CRISTINA: The way it works is in the so-called "library," you have these different notebooks, and they have one owner, the administrator of the business library. Independently of whoever is joining or leaving, those notebooks stay the same. Everything one person saves into that note will stay there, independently of that person coming or going. If you come up with the right structure from the very beginning, with, say, an HR notebook, everything related to HR issues stays there.
>
> For bigger companies, we have the expert's name show up whenever you're looking for a topic. Evernote will indicate the person inside the company who really

knows everything about that topic. That helps, because if you don't really find what you need, then you know whom to contact.

Monika, you have an amazing culture, with a flat organization, without power queues. How do you actually solve the decision dilemma? Who decides, the individual or the group?

MONIKA: I was waiting for that question.

For projects, the project team always decides in whatever way they see fit. For strategic decisions that are important for the whole company, we all decide.

It's not a top-down decision-making system, and it's also not a democracy. It's a Sociocracy[121] or Holacracy[122]. It's a newer term. The difference is that the number of opinions doesn't count, but rather the strength of opinions.

We sit together, and somebody brings up a certain decision. That's the first difference, maybe. You don't just start off with a random topic, but you always have a decision as the basis.

Then it gets very technical. We ask: does everybody know everything to really make the decision? Then, everybody has been heard. Then, we actually take the decision if nobody is against it.

In essence, every one of us has a veto right, but you have to think about it as an emergency brake. You only use it when you think: we as a company, we can't do

121   Sociocracy is a system of governance that applies the principles and methods of sociology, cybernetics, and general management theory. It uses systematic observation, measurement, and experiment to test the success of its decisions and its ability to achieve an organization's aims.

122   Holacracy is a system of flat organizational governance trademarked by HolacracyOne, in which authority and decision-making are distributed throughout a holarchy of self-organizing teams rather than being vested in a management hierarchy.

that. I have to get off the train if we take that decision. That's something that you have to train a little bit. Is it really that important for me personally? Is it just something that I think I don't really like?

Over time, you develop a feeling for your personal limits. If somebody draws that veto card, we just don't take the decision. Just as with an emergency brake, you get punished for that. You have to sit down with the persons who initially brought that decision, and figure out a new solution. You have to come back to the meeting next week, or next day.

When you have flexibility, it brings choice, and choice means freedom. However, the more choice we have, the more it paralyzes us.

At some point, especially in technology, there's so much to choose, that you eventually go back to pen and paper. It gets you back to basics and gives you some sort of feeling for what you do.

You can do whatever you want. You can do a side project. There's no destination offered. You are flooded with information. How do you feel that is going to change in the future, and how can you make decisions in a world where you're just overwhelmed by the amount of choices?

CRISTINA: That's a very good point. Back to pen and paper.

This is really funny. A couple of years ago, many people were saying Evernote's biggest competitor was PostIt, because people take notes on Poslit and Moleskine and such. For us we thought, why?

This is not really the case, because we are very humble about how we look at our users. We don't want to kill paper in any way. On the contrary, if you're using pen and paper, we want to figure out a way to support that. We've had these partnerships. We've launched with PostIt and with Moleskine. If you want to give digital life to your physical iconic objects, then you have the possibility to do so.

It goes back to this need of us finding the new way of living and working. It's not easy. This is the beauty of our situation. We have so many possibilities. It's a luxury situation at the end of the day, because it's a good problem to have, just to be able to choose. It goes back to spending time with yourself, which is not easy. It involves really looking at what are you trying to do with your life. What are you trying to do here? What is your goal?

How can you use whatever resources you see around you, either technology or people? Networks, people who can support your growth? Things to learn that can be various shapes and forms?

> You have to first learn to listen to yourself, and to also realize that it is absolutely okay to change, to have different goals as you grow. <

We have to learn to be more mindful and more honest with ourselves. We cannot take life for granted and say, "I'm just going to do my thing every day. Get a job and try to get my salary." There are so many unhappy people doing this. For the very first time, we have the possibility to change this. You don't have to be unhappy for the rest of your life.

It's not easy. You have to first learn to listen to yourself, and to also realize that it is absolutely okay to change, to have different goals as you grow, to see life in different ways, and to try to adapt at each step. You must look around you and see what could help you grow. Is it technology, is it people, is it learning something new, is it trying something new? It is not the easiest spot, but I think it's what brings, at the end of the day, the most meaning to our lives.

I'm following all the developments with Zappos, overseas, so I'm curious to hear from you, working in that setup. How scalable do you think it is? At some point in growth and decision-making, there will be some tension, right?

MONIKA: That's actually one of the questions we are dealing with at Dark Horse right now. It is scalable, but in a different way. You have to think of individuals sort of like cells, or individual teams, who are just combined by a single mission or vision or just a single brand. They are not centralized. That's a different mode of scaling. If you bring 150 people into one room, that sort of decision-making would tag on forever.

We realize that not everybody needs to be involved in every decision. Everybody's interested in every decision.

What is your personal view on the switch that Zappos made, moving very radically towards Holacracy?

MONIKA: Last year, Zappos switched to Holacracy, a similar decision-making process, and a similar governance, with elements of the governance we have at Dark Horse.

I have to say that I'm a little bit sceptical of the boss getting rid of himself or herself. What we see from the press is that it seems to work okay. But this also worked because they asked everybody in advance if they wanted to be part of it. I think 20% of the people chose not be part of it, so they were let go. Obviously, you have to be willing to give up some sort of aspirations. It's not possible to be the boss of your team anymore, because that doesn't exist in the system. It's not for everybody.

I think we'll round up. From what we've heard so far, it's not about the hours, it's about the quality of what we do or do not call work.

I think at Dark Horse, it's about fulfilment of life without any limitations. It feels like your work has become your life, and I'm very impressed to see how you champion that.

We discuss this because we feel the knowledge workers do need new ways to unite. How can knowledge workers in the future be more empowered, collaborate, and create a better future?

> MONIKA: There's not one cookie cutter way, saying, "This is how you have to do it, or this is how it works." We can't say, "This is how we did it at Dark Horse; here is your recipe. Do it wherever you are."

>You have to adapt your work to your life and not your life to your work.<

You really have to think about what you want to do. That's hard work. Coming up with our solution for our needs was hard work, and it still is. The basic principle is that you have to adapt your work to your life and not your life to your work. I think if you keep that at heart, that's your solution.

CRISTINA: Shouldn't we be more concerned about actual society, and what we are doing altogether? We shouldn't necessarily keep up with technology development, but create a new model that allows us to work in a smarter way, to be happier, and to contribute to a distribution of happiness and resources around the world.

Especially now with the technology we have on hand, social media and so on, more than ever it's possible to come together and find a new way of policy making that allows for social entrepreneurship, and ideas that can help us find a new way of living.

It's not so much that technology is going to take away jobs or destroy our future. It's really in our power to actually do something about it today and try to use whatever resources we have on-hand to come up with something smarter that benefits us all. If you ask, I think the majority of the population of this earth would be in favour of coming up with systems and models of living and working where you wouldn't feel unhappy again. This would be instead of going to a lousy job for 40 years, just because you have to.

With this new smart leadership, you can figure out ways in which technology can create new jobs that can benefit us all. It requires a lot of effort, us coming together, discussing, and figuring out, and it's not easy. The easiest part is just to outsource to technology, letting technology do everything. We'll just sit down and relax.

It's important for us to voice concerns, and it's equally important to challenge what we have today, to ask ourselves if we can do better.

> Can we live smarter?
We do have the power
to change. <

5
Disruptive Powers and the Misfit Economy

Marek Tuszynski
Alexa Clay

PANEL DISCUSSION, BERLIN, MARCH 14, 2016

As political leaders fail to create a sustainable future, social innovation manifests elsewhere. Deconstructing the world and re-building it is not the task of the political elite, but rather one for new visionaries who take on global challenges together.

As one of the misfits or outliers working on disruptive innovation, what strategies and tactics are key to mainstreaming your ideas? When is the right time to work within the system, and when should you spinout? Understanding your stance and your theory of change is critical. Are you a reformer working on the inside of our institutions or an explorer on the edges of the current paradigm? Are you a bohemian, building a singularly new world? Are you an activist shocking and provoking our existing systems into greater awareness?

Our relationship to the system determines our identity and legacy. We're born into cultures that we never asked for, that we didn't shape. Many of these cultures have pathologies and inherited command-and-control structures that are no longer relevant to the future we wish to create. As a result, we find ourselves moving through the world as culture hackers. Frustrated by pre-existing cultures, we hack at dysfunctions; we create the possibility for new worlds.

But beyond a particular project or technology, what is the culture, the way of being, the consciousness that we're using to infect the world? How do we steward that culture and embed it into the "prototype"? In the midst of massive economic and cultural transition, how do we re-program the institutions around us?

We discuss new roles and responsibilities in the world of tomorrow with Alexa Clay and Marek Tuszynski, two leading social innovators and entrepreneurs, who challenge the present, outline a collaborative future and inspire holistic leadership.

PANELLISTS

ALEXA CLAY is a writer, culture hacker and innovation strategist. She is the co-author of *The Misfit Economy*. A leading expert on subculture and innovation from unlikely places, Alexa is the Co-Founder of the League of Intrapreneurs—a movement designed to spark corporate social revolution. She is currently powering up a movement for a do-it-yourself economy across small town America in partnership with the BLK SHP. As a steward of Wisdom Hackers, Alexa has built a global community that delves into life's burning questions. She is Co-Founder and Director of The Human Agency, a collective working to incubate breakthrough ideas for humanity. Previously, Alexa pioneered work at Ashoka, focusing on new economy, job creation, and scaling social innovation. She has also led research and analysis on social innovation models in the health, energy and finance sectors and worked with leading Fortune 500 companies on sustainability and innovation strategies.

MAREK TUSZYNSKI is the Creative Director and Co-Founder of Tactical Tech, an NGO focusing on advancing the activist community through the use of information and digital technologies. Marek is a restless producer of creative and social interventions that span across various media: radio, television and Internet, along with workshops, book sprints and endless conversations. Activism, innovation and creativity are the major driving forces in his work, ultimately promoting marginalized voices, opinions and worldviews. Previously, he co-founded The Second Hand Bank, as well as the International Contemporary Art Network. Currently, his focus is on producing interactive and static visualizations that represent complex social and political issues. Very recently, he produced and directed a series of documentary films for

Tactical Tech entitled *Exposing the Invisible.*

Alexa, your book is setting a tone. It begins with a discussion about creativity from pirates, hackers, gangsters and other informal entrepreneurs. You say in your book that we can learn from those people as much as we can from Steve Jobs and Elon Musk. That's quite a statement.

How do you characterize misfits? What are the main perspectives or aspects?

> ALEXA: Initially, the book started as a joke project. I gave a fake talk at SXSW[123], because they were really annoyed with this idea of disruption and innovation. I wanted to say, "Well, what if the next kind of social media trends we can learn from Mexican drug cartels? Or, what can we learn from financing techniques that terrorists are using?" It was kind of this tongue in cheek irony.
>
> The more research we did, the more we actually looked at some of these personalities and realized there's a lot we can learn from the underground, informal, and black market economy. Specifically, we were really trying to transform this persona or this script around entrepreneurship. So often, the idea of an entrepreneur that comes to mind is this Silicon Valley-Wunderkind. We looked at this idea of the entrepreneur as the gangster, the entrepreneur as the hacker or the artist or the activist.
>
> When I first started using the word misfit, whenever I told my mother my idea, she sort of made this face. She's like, "Misfit, what do you mean? Do you mean weirdos? What does that mean, exactly?"
>
> Over time, we looked at four different typologies of misfit actors. The first was those underground sets: people who are really marginalized from the mainstream economy, who are really in those grey and black markets.

123   South by Southwest is an annual conglomerate of film, interactive media, and music festivals and conferences that takes place in mid-March in Austin, Texas.

We also looked at the runaway misfit: people who are voluntarily choosing to drop out of the system, somehow. There's this question of whether or not you can actually ever drop out of the system. I spoke to a lot of Hermits who, interestingly enough, have their own electronic newsletters for staying connected. That was one personality, the kind of Bohemian identity.

We also looked at people who are poking and prodding the system: the activists, people who are questioning the system from the outside.

Lastly, we looked at this idea of the entrepreneur, the real "insider misfit"; someone who has a counter cultural agenda but is a bit more camouflaged with that agenda. Or, people who are actively trying to transform bureaucratic cultures from within; people who are building out future business models, which are trying to affect change at scale. I think a lot of them are these Robin Hood types.

You also see saboteurs within those institutions: people who really believe in a theory of change that is all about taking down the existing system.

Part of the fun in the book was actually seeing the many hacker tactics that were used by entrepreneurs, for example, or how artists were actually developing their entrepreneurial mind-set to deal with this transition economy. Some of the fun was in seeing those interconnections.

It was also interesting, in the book, that you mentioned some principles that make you a misfit. Do you have some examples? What can we learn from the hackers and the hustlers?

ALEXA: The book is organized around five main principles.

The first is this idea of *hustle*, which is something you see within the black market economy; it's the street

hustle you see within Silicon Valley. It's that resourceful-
ness, frugality, being able to do something with a little,
the sort of dodge determination to get things done, that
burning drive and ambition.

>This copy or open-source culture is influencing a lot of these misfits.<

We also focused on hacking: hacking not just as computer hacking, but also as culture hacking. I think a lot of
work and my vocational identity is centred more around
culture hacking. How do we transform cultures? How
do we not just see cultures as stagnant things to study?
How do we transform corporate cultures? How do we
transform norms around money, around relationships?

Then, we looked at the spirit of copying. This idea that
actually, open source culture is really influencing a lot of
these misfits. How do we think about ownership within a
very different paradigm?

In many ways, I was raised by misfits. My mother studied
alien abduction. She was talking to people around the
world who thought they'd been abducted by aliens,
which was very scary to my childhood self. We were
talking about spaceship earth, and part of me was
always like, "Well, I want this alien conscious." I'm more
interested in being this ambassador and was really dis-
appointed over time. I have an egoic-level that I wasn't
being abducted.

All to say, I think there was something about my upbringing that had me really question the western consciousness that we were born into. I think that's the spirit of
provocation. You can actually see limitations within consciousness, and also begin to see alternative realities.

One of the subcultures I've written a lot about recently is Live Action Role Playing (LARP). Maybe you associate this with, like, nerds running in the woods or in Berlin with swords in parks. Sometimes that's true, but it's also, I think, a much more sophisticated genre, particularly if you're a Nordic LARPer.

The whole tradition there with LARP [124] is you can actually prototype futures. You can live out any kind of reality that you want to. From the perspective of provoking, it's amazing to actually be able to pop these temporal realities and play a character, with these design constraints. We were at the POC21 eco-hacking camp [125] together for five weeks. I know you'll share more.

Yes, indeed. That was actually the last time when we saw each other; in France, at the POC21 camp.

ALEXA: We were in this camp, using compost toilets and spending five weeks in this crazy situation. We were kind of in an extended LARP. Everyone left that experience feeling like they had this hunger for living in community in a very different way. We were living and producing goods in a very different way. Our identities at that camp weren't as consumers. In this temporary experience, we were able to hack our actual economic identities.

> In this temporary experience,
we were able to hack our
actual economic identities. <

124   Live action role-playing games (LARP) is a form of role-playing game where the participants physically act out their characters' actions. The players pursue goals within a fictional setting represented by the real world while interacting with each other in character.

125   POC21 is an international innovation community on sustainable living that started as an innovation camp. The camp brought together 100+ makers, designers, engineers, scientists and geeks. Their goal is to overcome destructive consumer culture and make open-source, sustainable products the new normal.

One of the beauties of it is that you have this phenomenon called "bleed," which is where the character that you play in a LARP starts to bleed into your own identity. It's a way of hacking yourself. That's really the spirit of provocation.

Then, the last principle is around pivoting, which I started calling rabbit hole for a really long time in the book. Our editor thought that sounded too druggy. It was sort of druggy.

It involved the idea of personal pivots that people make. Almost all of the misfits that we profiled have this enormous ability to jump from one world to the next, to bring one discipline into solving a problem that someone wouldn't have thought of before. That's roughly how it's organized.

Marek, with Tactical Tech[126], you were challenging political systems.

You might be a misfit in your own way, regarding what your approaches are. What is your perspective on the role of misfits?

MAREK: I would redefine the term of the misfit in the first place. I think in the book, this is fantastically described. I've read 78% of the book, and it's a great 78% so far.

> In the environment I work in, there's very little that you can do officially or legally. <

You referred to me not as an individual but as an organization. I will speak as organization on this point.

126 *Tactical Technology Collective is a Berlin-based non-profit that explores the political and social role of technology in our lives. It works at the intersection of tech, activism, and politics and reaches three million people worldwide, each year, through events, training, and online resources.*

You can approach me individually later, if you want. Misfits, for us, is kind of when you use the creative part of your brain, not against something or for something specifically. It often happens when there's a very scarce set of opportunities for you. You have to come up with an idea that would go against the stream, or you have to approach a problem sideways. Often, in the environment I work in, there's very little that you can do officially or legally.

Our work is often focusing on places. These are not necessarily geographical places. They're places where people are not able to express themselves. They're not able to experience the fundamental rights such as accessing information or creating knowledge or organizing themselves. There are a lot of these people here, not in this room, but in this city of Berlin.

For us, the power of information technology is that maybe for a second, it may enable a lot of people to do things they otherwise wouldn't be able to do. We are not very romantic about that either. It has pros and cons.

In your work, you probably run into the definition of what is legal and illegal very often. Through providing these tool kits and guides to activists and regions where it is quite difficult to operate, what is your personal definition of legal and illegal?

MAREK: It's interesting. I won't answer that directly. The answer is kind of a true story.

On one hand, what is legal and illegal is determined by whoever is in power. That can be a power in the house. In your household, it can be who has the remote control. What is normative in a society is more complicated.

For us, we often face a situation that we consider very legal, coming from our upbringing. When you look at the design of the issue, then it's much more complicated.

I will differentiate things into illegal and non-legal, if you like. There are moments in your life when you can tweak, you can push the system where you're still within the normative framework of not being criminal. Yet, you are challenging those who set the norm.

In other cases, you have to be head-on illegal, and you have to do something that would be considered criminal. Then there's kind of a fine line.

If Facebook is paying 4000 pounds tax in the UK, is this illegal? Is this legal?

You have this notion, in different places, where, if actors are big enough, either heads of states or corporations, they can push norms much further than individuals. Often, we punish individuals, not the large institutions.

> What is legal or illegal is determined by whoever is in power. <

I would like to look at these problems of legality and illegality in the normativeness in the context between individual and institution and also the state.

That's a perfect answer. It's questioning what the framework is.

Alexa, I'm sure when you were talking to all those interviewees, your concept of that might have changed as well.

ALEXA: I would actually agree with your point. The boundaries around illegality are always changing. Who is determining what is legal and illegal?

There were great examples within this idea of copying culture. The whole U.S. industrialization was paved, in part, because of patterns that they stole from Europe and then commercialized entirely illegally in America. It's a bit ironic that the U.S. is so focused on patent protection, particularly against other countries: Brazil and India and definitely China. Now that it's beginning to export its own products, even China is beginning to develop a different orientation to patents.

In so many ways, the history of capitalism is also the history of these shadow markets. During periods of rapid industrialization, you always had these enormous, smuggling markets.

We look very kindly on some of these historic rogues or historic criminals. We have a lot of empathy when we read books about them or we see television programs about them. Then, in the present day, somehow that empathy goes away. Somehow, in the present, you see things more through this lens of good and evil. It takes that historic memory to look at things with a little bit more perspective.

I think we've gotten some slack with the book, around this question of, "Are you glamorizing illicit activity?" Certainly not. I think there are tons of things you can read about how horrible criminals are. I think our perspective was: we will show how a lot of people who were born into very different kinds of circumstances and poverty are creating products and services and operating in slums and creating a livelihood out of nothing. We're looking at things through the lens of creativity and innovation. We're not really bringing a moral lens to things.

> You can be a cog
in the machine and work
for a Mexican drug cartel
or the mafia. <

At the same time, I was interested in characters that were born into hacking cultures. You can be a cog in the machine and work for a Mexican drug cartel or the mafia. For me, those weren't misfits within those organizations. Those were conformists within those organizations. I really tried to focus on people who are transforming cultures. Someone like King Tone[127].

King Tone was the former leader of the Latin Kings[128], one of the biggest Hispanic street gangs that originated in Chicago. He tried to do an entire change management approach for a gang organization. He said, "How do we transition this underground gang into a social movement?"—like the Black Panthers[129].

A lot of people, the FBI included, accused him of running a vicious PR campaign, sort of greenwashing the gang. He was making concrete changes. He started a task force within this gang when another person was running it to really figure out how they could develop activist skills within this gang organization. How could they use the gang as a feeder, as an important labour incubator for connecting young people from low-income communities with necessary job skills? I think that was really incredible. He was shut down because he was very deeply threatening to the powers that existed.

There is this bigger question of, like, how insurgent should you be? I think as a social innovator, my cultural hacking is fairly tame, compared to more radical approaches. In certain ways, that means it can be more successful. I think they are all different kinds of personalities that can bring different levels of annoyance into

127   Antonio Fernandez, also known as King Tone, is the former head of the Latin Kings. In 1999, he was sentenced to 12 to 15 years in prison for selling narcotics.

128   The Almighty Latin King and Queen Nation is the oldest and largest Hispanic street gang in both the United States and worldwide. Its roots date to 1954 in Chicago, Illinois.

129   The Black Panther or the BPP was a revolutionary Black Nationalist and socialist organization founded by Bobby Seale and Huey Newton in October 1966.

systems. Really being able to understand the choreography between those actors becomes an important part of social change.

Marek, I read about how you supported environmental activists in Africa by providing tools and advice on digital media usage. How has that impacted their role, their campaigning, and their ability to reach out and make a change?

MAREK: This is the most endangered group of activists around the world. The most killings are against activists or investigative journalists working as environmental activists.

We did an iteration of a tool kit that we've been running for 10 years. It's helping activists to protect themselves, protect their information and their sources. The process of trying to create change is usually very long. It's not three days, it's not three weeks, and it's not even three years. It's very important to preserve the power and abilities of people, and also give them tools to work with information and evidence.

The technology is just a medium and is changing every moment. People are using technology, and for us, they create change. Again, and kind of in contrast to the book I think, our focus is not necessarily on the individual. We address the people who aren't the story, not the entre-preneurs. They don't give their name to the work, be-cause they can't or don't want to. That's why we publish our materials to enable anybody in any way to access them as anonymously as possible. So far, we also do not collect fantastic stories about how what we offer is being used. We don't necessarily want to know that.

You work on two layers. One layer is this securing the privacy of the people you work with. The other layer is also visualization and making the data visible. You're working with the invisible and making it accessible to a broad audience.

MAREK: The real currency of any kind of activism and of rebellion is information; that's what people have, nothing else. This is the way you can challenge yourself as a society or power, if there's any power that you can actually see.

For us, the evidence that information is data is that we talk so much about innocent ears. Having it is one thing, and doing something with it is another thing. Obviously, there's no leasing of visualization or a poster or creative act that would change the world or even a little fragment of it. What it does is it enables people to recognize that there may be a problem that they can act together to solve. For us, these ways of working with data information and evidence is more finding ways of collaborating and understanding the world.

Since we now have powerful devices, we analyse information as never before. We can also translate narratives and propose these narratives to channels that were not available before. This is what the work is about. Using a combination of visual storytelling and knowing who you're trying to influence with that information.

In line with the narrative of the book, the biggest challenge we have in the group is not necessarily the specific persona of a power, wherever it is, but those who are neutral, indifferent, and interested. I think convincing people who avoid difficulties is the most powerful way of doing activism.

At your exhibition, Nervous Systems[130], at the Haus der Kulturen der Welt (HKW)[131], you created interesting installations on data

130 Nervous Systems was an exhibition, focusing on Quantified Life and the Social Question, holding artistic works by Vito Acconci, Timo Arnall, Mari Bastashevski, amongst others. It delves into concepts, such as, "Can our inner thoughts be transmitted by our eye movements? Can our future actions be predicted by our current behaviour?"

131 Haus der Kulturen der Welt, or House of the World's Cultures, is located in Berlin. It's Germany's national centre for the presentation

visualization. What do you think we can learn from big data? How to reorganize systems and empower activists?

> MAREK: Yes, the exhibition is, in a way, about big data. There are a lot of exhibitions about big data in Europe right now. We try to be different. We try to look at data through the human perspective. When you enter the space, you see everything from the human eyesight view. You see humans in images, in historical examples, and in some current political examples. For us, that's critical.

> The data is mesmerizing. It's so big that you can't compute it. You imagine that it would give you some answers to very important questions. We look at two aspects. On one hand, you have people who believe that if you gather all the information possible, then you can stop the disease, you can stop the inequality, or you can look at spaceship earth and move it in another direction. Hopefully, it won't do that.

> On the other hand, we struggle with the fact that the software, the code, the algorithms to analyse this data are written by us, and we're not very perfect. The analysis we do and the patterns we find are not necessary the ones that are very helpful.

> When you talk about data and politics, often you talk about the dystopian view: we're living in the world of big brother, surrounded by surveillance. Our take is more that we're actually living in a data society now that is more about supervision.

> You have this fascinating, more critical, utopian view that we're living in a post-privacy world, so we should take more care about our data. Partially, there is truth in it, and that we can do that.

and discussion of international contemporary arts, with a special focus on non-European cultures and societies.

Mark Zuckerberg said we're living in a post-privacy world. This means: give up your privacy and use Facebook. A year later, he built a house and four houses around his house for 40 million dollars to protect his privacy.

We're also trying to say that instead of disappearing, privacy is gaining new value. We're looking at who can afford it, why, when and so forth.

At the exhibition, the Sickweather App challenged the well-known, predominant narrative of the Silicon Valley entrepreneurship and what a successful entrepreneur is.

Maybe you have something to say about this?

MAREK: At the exhibition, the Sickweather and HealthyDay App are examples of representative applications for specific entrepreneurs in Silicon Valley. I have to explain what the Sickweather application does. You run it on a mobile phone or a computer. You can see who is sick around you. You can see whether they have the flu or any kind of illnesses. The data is gathered centrally.

HealthyDay are a subsidiary of Johnson&Johnson; they developed the application. The apps offer services to governments on when to provide health support and to give insights on what's going on in the country. The data is not only taken from you, you also receive a service. When you intentionally write into the app, "Oh, I have a headache today," that information is sold to third parties via Facebook or another service. Whenever you use words like stomachache, this goes immediately to Sickweather. That enables them to create their service.

Our take is that there's nothing wrong, theoretically, in creating a sophisticated surveying system. On one hand, you feel very empowered by it because you know which neighbours you should avoid. But you are the dot in this map as well, and other people will avoid you when you are sick.

When you take that to the political level, it becomes more complicated. Maybe it becomes about deciding your credit score. Then, we are living in a more complex environment. We look at the entrepreneurship of Silicon Valley through large corporations like Google, or Facebook, from this point of view. Some of them, not all of them, have probably the biggest aggregation of our knowledge in the history of civilization. Nobody knows more than Google about you. They know more than your most intimate friends, because you ask Google questions you wouldn't ask them.

> Nobody knows more about you than Google. They know more than your most intimate friends. <

They also aggregate the biggest wealth. They became the wealthiest corporation in the world in February, and who is their competition? Apple. The old idea of corporations is gone. It's not Shell. Yet, if you look at the revolving door and access to power, they can look at elections and other things differently. That's very unusual.

When you talk about entrepreneurship, especially data-driven entrepreneurship, it's not straightforward. It's not something that I would be proud of, now, from a political point of view. Yes, it's driven by massive creativity, but who can challenge that power?

Alexa, that's exactly what you can give us insight on with your misfit strategies. How can these misfit strategies change or influence mainstream phenomenon or corporations of that size, speaking about social innovation and corporations?

ALEXA: I think you see misfits' strategies operating on all different scales.

For example, a little performance character that I've been doing for a few years now is the Amish futurist. She's an alter ego I use where I dress up like as an Amish woman, I wear this bonnet, and I go to like Tech Open Air (TOA) or SXSW. I ask start-up founders these really hard questions. Why are you developing that app? What is its bigger purpose in the world? That's kind of like performance art meets Socratic inquiry.

> It's performance art meets Socratic inquiry. <

It's interesting to hear that actually, so many of these people are trapped within the Silicon Valley mold. When they articulate what they actually stand for, the values that they stand for are so different from what they're actually doing in the world. A lot of people who work for huge companies check themselves at the door. They're not allowed to bring all of themselves into the workplace.

What interests me more are characters like entrepreneurs, people who are able to step into a company like BP or Ford and speak truth to power. People who are trying to use those resources in clever ways. I think it's good to be critically minded of these companies. Some of them won't actually be able to transition. I think government has to do a really good job of regulating and restricting licence to operate.

In San Francisco right now, all these companies are privatizing resources, rather than building city infrastructure, rather than putting their taxes into transport links. They have their own private buses. That's really dangerous, when you don't see companies contributing to a core means, in the ways that they used to. But there are entrepreneurs in government within bureaucracies. There was a guy I met who worked at Ford Motor Company who pioneered human rights practices within the company. He got the company thinking about producing things beyond just automobiles.

There's maybe a misfit addiction, sometimes, to falling into this trap, where you just want to be the lone, mis-understood outsider. I even have an addiction to outsider perspective. I love coming to places and figuring it out, rather than feeling part of cultures I find gross.

Are there examples of misfit ventures that became mainstream? Shouldn't the goal be to really design a new normal?

MAREK: Before Alexa answers that question, I would actually like to challenge it.

I say that all the people who funded Google are the misfits. It seems like the face of capitalism they're pro-moting is based on a principle. They took over the role of the avant-garde and misfits and they have become creative pushers of the system.

> Google is pushing every behaviour to the maximum, until somebody tells them it's not okay. <

What Google is doing, for example, is pushing every behaviour to the maximum, until somebody tells them it's not okay. Then, they value it against how much it's going to cost them to get lawyers on that case. Those are the ethics they use. This is how they use legal frameworks. In that pure sense, yes, they are the real pirates. It's so difficult to challenge them, because they are rich.

In a way, that creates a strange environment. They are how you always wanted us to be. On the other hand, it's only just a few of them, and they are driven by a very specific mind-set that believes in the value of data and information. It's immediately controlled. We don't see that control yet, per se. It's very strange to talk about it, because whatever they do is also empowering us. We've never had such good ways of translating one language to another. We've never had better maps. We've never had better email. You name it; we could never find things better than we do now.

In a short run, they are fantastic. In the long run, they create an environment that we'll be living in very soon. We probably already are, which is very hard to challenge.

And actually my question to you would be: what kind of misfits do we need to challenge that system that these other misfits created?

ALEXA: I think another question is around mission drifts. A lot of things that emanated out of San Francisco in the '80s and '90s that were precursors to the companies that you see now were actually really counter-cultural ideas. People thought the Internet would be radically different than it is now. I think a lot of people felt betrayed by the dystopia that it became. You had things like the Whole Earth Catalog[132] and festivals like Burning Man[133]. There was this real awakening to the power of citizen production and this idea of disrupting companies, altogether.

Then people sold out on those dreams for various reasons. One of the main reasons was finance.

If there is a sector where we need more misfit energy, it's really within the financial system. How do we create

132   The Whole Earth Catalog (WEC) was an American counterculture magazine and product catalogue published by Stewart Brand several times a year between 1968 and 1972. The magazine's focus was on self-sufficiency, ecology, alternative education, DIY, and holism.

133   Burning Man is an annual gathering that takes place at Black Rock City—a temporary city erected in the Black Rock Desert in Nevada. It's an experiment in community and art, influenced by 10 main principles: "radical" inclusion, self-reliance and self-expression, community cooperation, civic responsibility, gifting, amongst others.

a more decentralized financial system? How do we train entrepreneurs, too, so that they don't have to sell out, so they don't have this mission drift? How do we make them better negotiators? How do we bring some of these founders who do feel deeply betrayed by the companies that they started back to this growing set?

> If there is a sector where
we need more misfit energy,
it's within the financial system. <

It's interesting to talk about ownership here, because one of your principles was copying. If we understand that this is one of the most fundamental processes in nature, to take what's there, to improve it, to measure it up, to reach the next level, then you can also say there is a kind of collaborative intelligence that improves our system.

Is the question of ownership a total myth? Progress happens when you open it up and copy systems, without patents?

ALEXA: This is how Tactical Tech is built underneath, right?

MAREK: Tactical Tech is built, yes, on the ethos of free software. I use the words "free software" on purpose here, because we commonly use open source. Open source only kind of addresses one of the aspects of what you can do with software, which is read the code and see if the code is good or what it's doing.

Free software is looking at the political aspects of coding. Free software actually enables you to become a political actor. That's our ethos. That was the very important beginning of our work. But when you look at property, it's a more complex question, because you probably wouldn't mind that some services are run and owned by specific institutions that were built by taxpayer money.

Rather, when you look at the finance system, you see we're living in a kind of a permanent crisis. It's especially a permanent crisis of the western world, or you could call it a permanent crisis of democracy. The economic crisis in the U.S. is massive. Sitting in Berlin, and especially here in the Soho House, we don't feel the crisis whatsoever. It's like, it's not here.

Can we create another financial system to do something with it? Is entrepreneurship the answer? I doubt it. I think this is one of the last waves, in a way, where we find the commodity that has an abstract value. If you look at how Google and other data gathering and data trading corporations are valued, it's interesting.

On the other hand, we know that we learn very little from this data. How were they valued so highly? And why do we have this drive of start-ups to join in the gold rush in California and San Francisco? It's interesting when you go to places, talk to people. You have this feeling of cowboys looking for the golden grains.

> Free software is looking at
the political aspects of coding.
It enables you to become
a political actor. <

There's definitely this rush of doing something totally out of the box. Marek, when you formed Tactical Tech, this was a group of people from all over the world that, at some point, decided the things you were doing were so complex that you should all be in one place. Then, you asked, why can't you have more from these sorts of initiatives? Not only free software but open source creative collective innovation.

MAREK: Tactical Tech was founded around political freedoms.

At the beginning, they were about the struggle between political powers and democratic countries and authoritarian environments and activists who were defending human rights.

At the beginning, we believed corporations would start using technology in a certain way, with encryption and open source. We thought that would create a more open environment.

Now, it's 14 years later. We find ourselves in kind of a double trouble. Our imaginary allies are probably much more problematic and complex enemies to our freedoms than we ever imagined. Yes, at some point, technology opened the ability to become political. Now, we feel like it's not there anymore. It's very complicated to explain that. Technology is neither good nor bad, but it is also never neutral, never.

How can you create something positive? For example, the blogging platform Medium[134] was a place for multifaceted opinions, alternative opinions, and challenging opinions. Now, we find out that it was bought by Alphabet Inc[135], so it belongs to Google. I was flabbergasted. Do I really want to work with Medium now, the way I worked with it before?

This also means that once you reach a certain level of relevance and success, you're sold. What comes back from these misfit tactics? What can you do to create something that is sustainable?

MAREK: The Internet is still what it was; it didn't change. What happened was, the colonization and kind of

134   Medium is an online publishing platform developed by Twitter co-founder Evan Williams and launched in August 2012. The platform is an example of social journalism, having a hybrid collection of amateur and professional people and publications.

135   Alphabet Inc. is an American multinational conglomerate created in a corporate restructuring on Cotber 2, 2015. It's the parent company of Google.

accumulation of information changed. Now, for many people in many parts of the world, the Internet is for instant Facebook, because that's the first layer on their mobile phone. They don't even realize that there's possibility for peer-to-peer, protected communication. You can pop up your own information-kiosk anywhere on the Internet, as you like, and you can do it in an open or closed way, however you like.

> The Internet didn't change; the colonization and accumulation of information changed. <

There are other ways for creating a trustful and confidential way of exchanging information. It's all built in the architecture of what we have. The problem is that it is becoming criminalized, now. On one hand, you are told that encryption is good when you use it, if you buy a Google service. But if you use Tor[136], then you are a terrorist. The technology is the same; there's no difference there whatsoever.

Let's talk about hacking. I like the term, actually.

John Draper[137] was one of the first hackers. He developed the Blue Box[138] and hacked the telecom system. While he

136 *Tor is a free software and browser enabling anonymous communication. It directs Internet traffic through a free, worldwide, volunteer overlay network consisting of more than seven thousand relays to conceal a user's location and usage from anyone conducting network surveillance or traffic analysis.*

137 *John Draper is an American computer programmer and former phone phreak. He is a legendary figure within the computer-programing world. He's long maintained a nomadic lifestyle.*

138 *A blue box is an electronic device that simulated an operator by generating the same tones employed by a telephone operator's dialling console to, for example, switch long-distance calls.*

went to jail, two of his best friends, Steve Jobs[139] and Steve Wozniak[140], started their first business out of the Blue Box. They improved it and sold it.

I think this is a wonderful example on how different the approach can be for hacking the system. How would you define hackers, and what are the motivations behind their actions?

ALEXA: Hacking is a real spectrum. There are all different kinds of hackers with very different types of agendas.

With the example that you just gave, you see constant elements of this. Something from the black economy then becomes either co-opted or mainstreamed. Napster, for example, paved the way for all these music file-sharing sites, video-streaming technologies. It was initially pioneered like an illicit pornography website. Even Shanzhai[141] copycatted cell phones, which then big phone operators started to use.

People have different orientations towards the system. Sometimes, it's about the system catching up. And what would be regulated 10 years ago becomes legal later on. I think the danger is when you have these incumbents like Google or other big companies that have the ability to define what that regulation looks like. They have the ability to potentially out-regulate competitors.

In the book, we looked at a wide variety of hackers. Anyone from a classic sort of person who believes in the freedom of information, to someone who believes in the

139   Steve Paul Jobs was an American entrepreneur, businessman, inventor and industrial designer. He was the co-founder, chairman and chief executive officer of Apple Inc.

140   Stephen Gary "Steve" Wozniak is an American inventor who co-founded Apple Inc. He's known as the pioneer of the personal computer revolution in the 1970s, alongside Steve Jobs.

141   Shanzhai refers to counterfeit consumer goods, including imitation and trademark infringing brands and/or particularly electronics, in China.

protection of privacy. Those things might sound counter-intuitive, but they are often mingled together.

There's social hacking; there's cultural hacking. Some hack a problem of violence. We profiled a reluctant misfit in the book who spent 15 years studying infectious disease. He worked with community health outreach workers, learned a model, and then used it to treat violence in urban cities. He developed a whole program for mediating violence. He created violence interrupters that involved informal police forces around communities. He did this based on simple insights of seeing the patterns of how violence spreads is exactly how infectious diseases spread. He was a hacker in the sense that he took one discipline and applied it in a very different context. Then, he faced a lot of push back.

Some of his hacks were based around this question of, "How do I get the system to adopt my approach?" This is a very different question than the hacker that believes in a more anarchistic kind of philosophy, which is maybe around, "How do I take down the system, or how do I build an alternative system?"

One of the metaphors that friends and I use a lot in Berlin is around the sort of *island* versus *peninsula* versus *mainland* orientation. Which is your approach to social change?

Some people are trying to do social change from the main land. They are the entrepreneurs. Other people are really in that hacker isolationist standpoint of trying to create an alternative kind of culture. They're trying to bring down the main land or disengage entirely from it. The in-between space is the peninsula, where you are attached to the main land, but you still have enough isolation, enough separateness to protect some of what you're trying to do. I guess if you're in the peninsula, the question is, how do you prevent some of the things you're working on from being co-opted?

Facebook and Mark Zuckerberg[142] now talk about themselves as a hacker company, which any hacker would object to. People now think that you can solve social problems within like a three-day hackathon. You've lost some of that insurgency. That's why we need people to be able to police those types of things.

> Facebook and Mark Zuckerberg now talk about themselves as a hacker company. People think you can solve social problems with a three-day hackathon. <

MAREK: I really like the way that you equate the misfit and the hacker. I think misfit is a nicer title; it's less scary, in a way. The hacker now means somebody that is dangerous, someone who will steal your data, your bank account and destroy your car while you're driving. This is very interesting, because, if you look at the *Webster*[143] definition from the time before media started describing hackers in a certain way, hacking meant improving, then kind of tweaking and fiddling, trying to make things better.

Human beings, by default, are hackers. We've taken technology from the early stages of Homo sapiens, if you like. It changed.

You also say in the book that Facebook's address, today, is 1 Hacker Way. Let's go back to the previous conversation about the power of these corporations and how they define their role in the society. They are the real hackers.

They're really pushing our understanding of how to organize society.

Most of them come from the single belief that this is very driven by cybernetics. If we only collected all the information and data, we not only can create artificial intelligence, but we can find all the answers to all the problems that nobody before, ever in history, was able to find out. That's quite scary.

Hackers don't always have the best image, but they can also be people who try to bring light to inequality or conspiracy.

How can they help to improve democratic processes or political structures?

MAREK: You're talking about two different things. I can't address conspiracy and other things in the same sentence, because conspiracy doesn't exist. Conspiracy is a surreal situation in which somebody is interpreting the world in a certain way. They think it's driven by specific groups or ideologies. The conspiracy theory is fictional, in most cases. There may be something behind it; there may be some truth.

Look at the problems a lot of hackers are trying to address. They work with data, and they try to turn evidence into something that is meaningful. They try to bring technology to places where there's not enough expertise on how to use it and what to do with it.

There are a number of examples, but are those examples really world changing? I would question that. I think they are often temporarily significant. But this is not how we can drive the ship called earth anywhere.

On the other hand, it's becoming very de-politicized. Some running hacker spaces are focusing on 3D printing from garbage plastic, which is quite sad.

Let's talk about "spaceship earth." Who is at the steering wheel?

You've mentioned these giant tech companies who influence our lives. At the same time, the hackers and hacker collectives, like Anonymous, are creating a counter-weight.

What is their role in where our planet is moving? What can their role be in the future?

> ALEXA: I think the hacker ethos is built around liberating information. That transparency push has been enormously successful. Anonymous recently did a campaign where they released the identities of KKK[144] members within the U.S. who were in support of Trump[145] rallies. I think they're providing a sort of moral policing of society. They're not waiting for permission.

> Part of the reason hacker groups like Anonymous[146] are so successful is they give people this anonymous yet collective voice. There is no boss of Anonymous. Anyone can participate in that culture, anyone can propose an action. It's very egalitarian in how it's structured. I think groups like that are really important as prodders, as people that are poking at culture and getting people to wake up.

> That awakening piece is huge. Some of the fascism rhetoric in the U.S. and in certain European countries

144   The Ku Klux Klan, commonly called the KKK, is the name of three distinct movements in the United States that have advocated extremist reactionary positions such as white supremacy, white nationalism, anti-immigration, amongst others.

145   Donald John Trump is the 45th President of the United States. Before entering politics, he was a businessman and television personality. For 45 years, he managed The Trump Organization, the real estate development firm founded by his paternal grandmother.

146   Anonymous is a loosely associated international network of activist and hacktivist entities. It became known for a series of well-publicized publicity stunts and distributed denial-of-service attacks on government, religious, and corporate websites.

in response to refugees is really scary. Who will be the people to steer the ship towards living out a different version of society? I think some will be grassroots communities, populous communities. Some of them will be hackers. Some of them will be entrepreneurs.

Amongst the change agents that I've worked with, I've learned it's really important for different people to have compassion for each other. The more empathy that we have for different people creating change within various parts of the system, the better.

I was part of the Occupy Movement[147] in the U.S. It's interesting to see how now, that entire vocabulary is the left progressive vocabulary in the election cycle. The language that Bernie Sanders[148] is using came from the Occupy Movement. It's now being used by Hillary Clinton[149], so she can spout her progressive ideals as well. Maybe that's the risk of co-option, that you have someone like Hillary Clinton, who is much more of an insider, gaming the system, using this Occupy language.

Actually, it's most powerful if you can spur this alternative power, if you can really invest in alternative power outside of the political process. This affords the necessary leverage for entrepreneurs to do their jobs.

147   The Occupy Movement is an international socio-political movement against social and economic inequality and lack of "real democracy" around the world. Its primary goal is to advance social and economic justice and new forms of democracy.

148   Bernard Sanders, known as Bernie, is an American politician who has been the junior United States Senator fro Vermont since 2007. He's a self-described democratic socialist, pro-labour, with an emphasis in reversing economic inequality.

149   Hillary Diane Rodham Clinton is an American politician who was the 67th United States Secretary of State from 2009 to 2013, U.S. Senator from New York from 2001 to 2009, First Lady of the United States from 1993 to 2001, and the Democratic Party's nominee for President of the United States in the 2016 election.

> Maybe that's the risk of
co-option, that you suddenly
have people like Hillary Clinton,
an insider, gaming the system, using
Occupy Movement language. <

I knew a woman who was working at a company that
used to leak information to Green Peace[150]. She was a
total entrepreneur. She said, "If you have this information, if you can do a campaign around this, then I can
get this through." I think that's the kind of ally-ship or
secret ally-ship that we have to be able to create.

MAREK: Disruption is extremely important, but not just
for the sake of disruption. Disruption is often important
because it opens the spectrum and gives the space for
people to exercise in different ways. It creates more of a
mental space, because it often shows that the emperor
is naked, or it challenges the way we see things. The
problem nowadays is that disruption is mediated through
technology. Technology turned out to be extremely good
at helping us to organize and mobilize.

We see waves of very strong organization, but we also
see that they die very quickly. The same technology
does not allow us to sustain these movements. The
reason for that is that, in a more complex, capitalist,
symbolic environment, technology is empowering us
as individuals. We have never been as individualistic
as we are while using technology.

This individual empowerment is taking away the empowerment of communities and groups across borders,

150   Greenpeace is a non-governmental environmental organization
with offices in over 40 countries and with an international coordinating
body in Amsterdam. Its goal is to "ensure the ability of the Earth to
nurture life in all its diversity."

across countries. We have these moments of empowerments, individually, as temporary groups like Anonymous or Occupy. But because we are so individually threatened by these devices, there's no way these movements can accumulate anywhere and turn into a kind of political proposition.

> Because we are so individually threatened by these devices, there's no way these movements can turn into political proposition. <

Occupy is a very interesting case. People were immediately there, in a single place, in a single location, trying to create a better world. They tried to create some rules, which had no chance of surviving, obviously. They had to leave that space at some point, and then technology betrayed them.

It's interesting to work in this environment. We have the notion that we're promoting technology. How do we make ourselves aware than we're living in this paradox of the self-environment, where we are more powerful when we are connected?

ALEXA: The same is true of politics as well. I think the biggest thing we have to overcome is this paradox of the self or this consumer orientation. The whole political system, at least in the U.S., treats each person as a consumer. All that you're asked to do politically or to further civic culture is to vote for someone else. It's an outsourcing culture.

Democracy is kind of about outsourcing. I think if you look at the great political movements of the past, it wasn't built on that outsourcing instinct. It's actually been based on a call to action that's much deeper, that's about collective mobilizing. It's about building common resources. It's about coming together and building resilient communities, often in the wake of government failure.

Technology is creating such a fragmented *attention economy*. It's creating so many conditions of further isolation and alienation. We don't necessarily have the capacity to engage in some of that collective work that must be done. Poverty is also a huge issue. People just don't have the time to engage in civic culture in the ways that they might want or should.

Let's discuss media. You say that in the past, there were smaller publications. Conspiracies were obscure.

Now, these obscure conspiracies become sort of news or commodity or believed in as if it were mainstream media, just because of the numbers.

How do you deal with that change in the media production and consumption?

ALEXA: It's a double-edged sword. The power of the Internet and new media means that you can have all these misfit communities that previously felt really alienated in their ideals. Now, they begin to create norms, begin to create consensus around some of their counter-cultural ways of thinking.

I was reading a book about today's witches earlier this morning. It was amazing. All of these witches learned about the Internet and could find each other online. They felt they were no longer these weird people, that they could actually share information and share spells with each other.

You see that also within political organizing. Suddenly, people with very different kinds of ideas can find their source of community. I think the other side to that is: the Internet is this crack machine in which you have so much traditional media being bastardized by this short-term attention span, with clickbait articles.

> The first wave of Internet media and clickbait is like the first wave of LSD. It's fucked us up. <

It's like the first wave of LSD. People thought they were beta testers of this new phenomenon. It really messed with people. I think that's kind of what the first wave of Internet media has done with us. It's really fucked us up emotionally and psychologically.

Hopefully, we're now at a point where we can be more intentional about the kind of media that we consume. In the same way that people have a preoccupation with nutrition and what goes in their bodies, we should be censoring the content that we consume. That's impacting how we see the world and what we're even able to think about.

MAREK: To expand this, I would say that this techno-logical environment in which we live also creates an environment of much higher precarity and lack of stabil-ity. When you don't feel like you know what's going to happen to you soon, you live in fear regarding the future. And fear is the kind of feeling that makes you start listening to and hearing different conspiracy theories and finding problems outside of your own known environment.

That's very nutritious for paranoia.

ALEXA: And for fascist leadership.

MAREK: Exactly, that's exactly how it comes to the fora. Maybe we should talk about witches. They're more fun than the fascists.

Marek, you discussed the dangers and the moral failings of big businesses, in particular, tech companies like Google and Facebook. You also touched briefly on the relative natures of legality and morality.

Are there big companies that have maintained their ethics? How do they maintain their goodness? What are the principles that they should strive for?

> MAREK: I don't think that evilness is in the company. I think Google maybe is evil, regardless of what they say. I think the evil part happens between the individual and the corporation itself. Google is so powerful, not because they pay us, but because they're using our data to make money. Our labour is free. We work for them from seven in the morning till midnight, but we don't feel like we're working. We feel like we're empowering ourselves.

> The most important paradox is that corporations cannot live without us, and we cannot live without them. <

> The evil exists between these two dimensions. Don't get me wrong, I'm not blaming corporations for how we use data and information and technology in a political way. I think the more important paradox is that they cannot live without us, and we cannot live without them.

> There's a number of businesses, some of them in Berlin, that try to have some code of conduct. They're experimenting with bringing ethics back to the fora. There are designers who think differently about designing products, for example.

> The problem is when you have big money and big responsibility with shareholders. By the end of the day, you have to deliver money. Other things become

secondary, tertiary. It will be interesting to see how we can look at the labour we do with recognized value from the beginning. We'll know what the value is, and we can invest in things with value. We don't have that opportunity. The system, now, creates this invisible bubble in the cloud, and then it disappears and appears somewhere else.

A lot of people are decent human beings, responsible individuals. Corporations can be like that, too. The problem with data-driven corporations is that it is easier not to see the consequences of using information.

Maybe even Google has management right now that supports that breach. Even if not, the point is: if we can create a structure allowing us to trust that the profit won't dictate the decisions completely, the long-term effects could be very powerful. It would be much harder to monetize and to build such structures.

MAREK: I agree with that. I think we're living in this problematic environment. Selling data drives the whole system. Until we come up with something else that would produce money for these corporations and create the infrastructure and devices, we are at a dead end. You can only apply software systems of hacking that parasite on the infrastructure they create.

Until we find another way, we're stuck with what we have. It's becoming more and more nasty, because there's more and more competition. There's more and more relationships between those who run this institutions and power.

ALEXA: It's very harmful when the abstract economy, this harvested data, becomes more valuable than the primary economy. Looking at the ratio of primary agricultural goods, like water and food and just basic needs, we need to assign a higher premium to these things and invest more energy into stewarding them.

What would be a path forward in terms of managing these resource constraints? Part of it is about shrinking parts of the economy that are unstable and speculative. It's about investing more energy into basic primary and secondary manufacturing economy. It's also about encouraging people to make things themselves, taking some corporations from some productions.

> It's very harmful when the abstract economy, this harvested data, becomes more valuable than the primary economy. <

Disruptive Powers and the Misfit Economy

6
Blockchain: Rewriting the Global Governance

Shermin Voshmgir
Vinay Gupta

PANEL DISCUSSION, BERLIN, MAY 4, 2016

17th century thinkers John Locke and Baron de Montesquieu provide the basis for our system of modern democratic governance. 600 years later, our society has evolved, but our constitutions still rely on Locke and de Montesquieu. Why is that?

Today, Blockchain is one of the most-discussed value exchange protocols. It has the power to transform the structure of our political and economical or resource-related systems on a fundamental level. Algorithms of present and future Blockchains have the potential to become the constitutional foundations for the structures of our future societies. And yet, so many people can't comprehend what the Blockchain technology means. With understanding comes imagination. This imagination can perhaps unshackle us from our current, stone-like constitutions and build a more versatile one.

For this discussion, Shermin Voshmgir and Vinay Gupta discuss the potential, the challenges and the feasibility of different Blockchain applications eventually governing our society.

PANELLISTS

SHERMIN VOSHMGIR is the founder of the BlockchainHub, a Berlin-based info hub and think tank focusing on topics like smart contracts, peer-to-peer economy and the future of democracy. She received her PhD in IT-Management at the Vienna University of Economics, where she once worked as an assistant professor and currently lectures Blockchain-related topics. Furthermore, she studied film and drama in Madrid. Her past work experience ranges from IT-consulting, Internet start-ups and filmmaking. Among others her films have screened in Cannes and at Documenta. Over the span of two years, she produced a monthly documentary series about democracy. She is also the founder of Lilon, an Internet-based travel company that offers app-supported, off-the-beaten-track trips along the Silk Road.

VINAY GUPTA is the release coordinator for Ethereum, a FOSS scriptable Blockchain platform. His main area of personal interest is using technology to end poverty globally, taking an engineering approach. This work has brought him through disaster relief (Hexayurt Project, an emergency shelter that was one of the first Open Hardware projects), critical infrastructure and state failure modelling (Simple Critical Infrastructure Maps) and cryptography/governance (CheapID). His work is in use for Burning Man (>2000 hexayurts constructed per year), the U.S. military (STAR-TIDES project) and with activists around the world.

VINAY: The Blockchain was first built as a backbone technology of the crypto currency Bitcoin, which is essentially a decentralized virtual money transaction tool with the capabilities of a pocket calculator. It can add, it can subtract, and it can store value. It supports a methodology to store and transfer data decentral within a network to exchange information directly peer to peer without any intermediary, like a bank. Through consensus of all nodes in the net, data will be transferred, stored, or deleted, through adding an information block in the data chain. Today this uses a lot of processing power, which makes Blockchain super slow.

I support a project called Ethereum[151]. We work on improving the scalability of Blockchain. In a couple of years, we expect it to be running 10s of thousands of transactions a second, like Visa today. The Blockchain itself is not fundamentally mysterious. It's just a completion of earlier stages of development in computer science.

In the 1970s, we invented the database. At this time in development, you had one computer per institution. In the 1990s, computer networks came mainly in the form of the web. Then, we went from one computer per institution to one computer per person. The software from the one computer per institution days didn't work very well for a one computer per person paradigm, where everything must be interconnected. This is the reason we still have these technical silos inside our organizations, where nothing can talk to anything else.

Let's say you have an insurance claim. You have to put your name and address into forms on the Internet 50 times, rather than doing it once. The Blockchain solves this problem of how to interconnect lots and lots and

151   Ethereum is an open-source, public, Blockchain-based distributed computing platform featuring smart contract functionality, which facilitates online contractual agreements.

lots of computers, so they have a common, sustained, reliable picture of reality.

SHERMIN: It's just a next step in the history of computing. We first had the computer, and then we had the Internet, where we connected the computers. We would have designed a very different set of rules on how to store and manage data and verify transactions, if the Internet would have existed before the computer. Blockchain is just a logical continuation of computing history. It's just the beginning.

In maybe two years, we might not even talk about Blockchain anymore. There will be an understanding of what we can be build on top of all these decentralized versions of Blockchain, like IPFS[152], Whisper[153], Swarm[154], etc. What will stay is the idea of decentralized data storage and data verification, no matter what we call it.

What are the different kinds of Blockchain and what are they developing towards? What will outlive the others?

SHERMIN: The first Blockchain was the Bitcoin Blockchain. Today, we are here talking about Blockchain, because first there was Bitcoin. There was the idea to have money without banks. Satoshi Nakamoto[155] thought of a concept of having money for business transactions, but without the central institution of the bank.

152   The IPFS, or the InterPlanetary File System, is a protocol designed to create a permanent and decentralized method of storing and sharing files. It's a content-addressable, peer-to-peer hypermedia distribution protocol. It was initially designed by Juan Benet.

153   Whisper is a part of the Ethereum P2P protocol suite that allows for messaging between users via the same network that the Blockchain runs on.

154   Swarm delivers shared equity with Blockchain technology for the sharing economy. It allows companies to incentivise these users with "shared equity," giving them competitive advantage.

155   Satoshi Nakamoto is the name used by the unknown person or persons who designed bitcoin and created its original reference implementation.

That was defined through Blockchains. We've been developing it for a few years now. The code is open source. The next step was that people took the Bitcoin source codes and created their own forks[156]. They modified the Bitcoin source code into so-called Altcoins[157] and tried to do other things.

One important step in this development was the creation of Mastercoins[158] and Colored Coins[159]. Users said, "Let's try to take Bitcoin as a token of transaction and build any kind of auto enforceable smart contract[160] into it." Vitalik Buterin[161] was also working on that very briefly, until he understood that it didn't make sense to do that. Then, he started to work on Ethereum[162] to program a whole new Blockchain. It's like a decentralized virtual machine where you can build any type of smart contract. It is much more flexible.

VINAY: There are basically two families of Blockchains: the stuff derived from Bitcoin and the stuff derived from Ethereum.

---

156   In software developing, project fork happens when developers take a copy of source code from one software package and start independent development on it, creating a distinct and separate piece of software.

157   Altcoins are cryptocurrencies other than Bitcoin. The majority of Altcoins are forks of Bitcoin with small uninteresting changes.

158   Omni, formerly called Mastercoin, is a digital currency and communications protocol built on the bitcoin Blockchain.

159   The term "Colored Coins" loosely describes a class of methods for representing and managing real world assets on top of the Bitcoin Blockchain.

160   Smart contracts are computer protocols intended to facilitate, erify, or enforce the negotiation or performance of a contract. They were first proposed by Nick Szabo in 1994.

161   Bitalik Buterin is a Russian programmer and writer primarily known as co-founder of Ethereum and co-founder of Bitcoin Magazine.

162   Ethereum is an open-source, public, Blockchain-based distributed computing platform featuring smart contract functionality, which facilitates online contractual agreements.

The big question today is how to make a Blockchain fast. The current trend is that systems are doing seven transactions a second for one Bitcoin, 20 for one Ethereum. These are basically toy systems. The critical question is even: can you make it fast on top of the existing technology platform? Is it possible to make Ethereum faster, or do you have to break it down and do something else?

We are not entirely sure yet. Some vastly smarter people are technically capable and working on it. They are confident they're going to make that happen. Then, there will be some number of years in which these systems either self-transcend or become obsolete.

But these systems are acting as storage of value—the "permanent" storage of value in the form of money or contracts. The constraints of how quickly that can evolve are essential. You can't just tear an existing Blockchain down and start over, because you have to maintain all the existing customer states, both money and contracts. This onward process of self-transcendence and migration is going to be worth an enormous amount of technical action and will happen intensely over the next five or ten years.

SHERMIN: I would like to cite Trent McConaghy[163], from ascribe.io[164] and BigchainDB[165]: "The first step was Bitcoin, and then Altcoin followed. Now, we have Ethereum."

Ethereum is the decentralized world computer, where you can build any type of decentralized application on top of it. The mentioned smart contracts or decentralized

---

163 Trent McConaghy is the Founder and CTO of BigchainDB and is working on a shared public database for the Internet.

164 ascribe.io is a service that empowers creators to truly own, securely share, and track the history of a digital work. Creators are given the tools to lock in authorship, set the intent, and gain visibility into where their work spreads.

165 BigchainDB is a decentralized database, beginning with a distributed database, adding Blockchain characteristics over the top.

applications are the next step, built on top of Ethereum. The next step, in particular, will be the DAOs[166], Decentralized Autonomous Organizations.

>Artificial Intelligence (AI) isn't futuristic anymore; it's already happening.<

Today, we are building the core level of Ethereum. We are still developing the basic protocols. And with this, we are building the first smart contracts and the first DAOs. The next step of development will probably be an AI DAO, a Decentralized Autonomous Organizations, based on Artificial Intelligence (AI). If you haven't read Asimov's Three Laws of Robotics[167], you should. Just a few years ago, they seemed so futuristic, but they're becoming really relevant today. AI is not futuristic anymore; it is already happening.

Combining AI with the Internet of Things[168] (IoT) and the Blockchain, we need to make sure we don't create the futuristic world of a Skynet-like reality, in which we live under technological super control. In a way, we are building Skynet[169]. This is why the discourse around Blockchain and how we build it is so essential today.

166   A decentralized autonomous organization (DAO) is an organization that is run through rules encoded as computer programs called smart contracts.

167   Three Laws of Robotics are a set of rules devised by the science fiction author Isaac Asimov.

168   The Internet of Things is the inter-networking of physical devices, vehicles, buildings, and other items embedded with electronics, software, sensors, and network connectivity, which enables these objects to collect and exchange data.

169   SKYNET is a program by the U.S. National Security Agency that performs machine-learning analysis on communications data to extract information about possible terror suspects.

How do you see Blockchain applied in education, voting, or operating businesses in the future? What are the next steps to apply Blockchain?

VINAY: There are conservative things we know will happen, and there are more radical things that might happen. The basic conservative level is still really, really shocking, since even today, none of the computer systems that operate our society work together. Just remember how often you have to put your address into a computer form, even within the same organization.

> There is still a pervasive brokenness about computers. We never figured out how to get the machines to work together. <

The internal system of one of the major banks stores 600 copies of a single customer's address. These copies must be updated when something changes. I told this to some other bank tech guys, and they said, "600, that's really efficient." That's conservative. There's a pervasive brokenness around everything that touches computers right now. We never figured out how to get the machines to work together.

The pervasive brokenness will gradually go away. I'm sure. There's this saying in England: "Nothing fades like the future, and nothing sticks like the past." Getting rid of the pervasive brokenness will probably involve just attending your job. But I'm sure that this will be chiseled all the way up, largely because of rising expectations and technology. It takes both factors.

The radical stuff, the hard-core line, is that we've no effective global government at all. We've seen the nation states flailing around for 20 years to get an agreement on carbon. They've done literally nothing. The ice caps are melting; we all know there's a huge problem on the way. There's the breakdown of agricultural systems in large parts of the earth, since it's uninhabitable and hot. The nation states do nothing. Even the progressive states do nothing.

Some countries have become largely carbon neutral, like Bhutan or Sweden. The majority of nation states are chewing on oil as fast as they can get it. The U.S. is trying new technologies to bring new oil to the table.

We need to use the power of the Blockchain, not just to constitute a decentralized banking system, but to constitute a global government that has the power to boss around nation states. If we don't assemble something at a global level, with the power to give the state orders, the state will simply destroy us through inaction. The climate processes are basically trade unions made up of 200 fat cats. Entities called nations sit around a table discussing who will pick up the slack when everything goes wrong.

> We need to use the power of the Blockchain, not just to constitute a decentralized banking system, but also to constitute a global government. <

Other countries, with major, dry, high grounds and stable climates, are telling everybody else, "Just wait until the disaster comes. Maybe we'll help you out."

The high ground here, if you want to talk about the potential of these technologies, is that we will institute global governance. We'll record the votes of individuals on Blockchains. No power in the world will have the ability to edit or pretend those votes never happened.

From this, we will gradually build a global democracy. We'll use statistical techniques to figure out when enough people have voted on something for it to be statistically significant. Does it take 1% of the population, does it take 5% of the population, or does it take 20% of the population? Of the entire planet, how many people have to vote on an opinion before it has to go away? This would depend on how devoted the opinion was.

The idea is that we should simply reconstitute democracy at global level, then push down on the states to make them obey. It's the really radical proposal on the table, and I think we should do it.

As mentioned, we need checks and balances. What do we do about the separation of power: legislative, executive, and judicative? Is it written within the Blockchain? Is it something that corrupts the Blockchain? How can we establish that separation of power in the future in the Blockchain?

SHERMIN: In computing, there's this notion of separation of duties. It's very interesting already in the Bitcoin and Ethereum community. This notion of separation of powers is translated into the separation of duties.

I'm really worried about the flexibility of these core constitutions, reflected in the Blockchain protocols. How will we keep up to date with future developments?

Is the separation of power in a decentralized system actually applicable at all?

VINAY: Very much so.

Blockchains are basically smart paper. If you write on it once, everybody in the world sees what you've written. Nobody can write anything, or pretend somebody else wrote something. Nobody can erase. You could pick any kind of governmental process and represent it many ways on Blockchain. There's no problem with building separation of powers if separation of powers is what we want.

I don't think there's anybody left standing who has actually done the kind of political architecture that marks that phase of political development. When was the last time a team sat down, looked at such technologies, looked at such government issues, then designed a government system?

I think Visa might be the last big example. Visa had really sophisticated governance arrangements in the 1970s or 1980s.They had incredibly sophisticated approaches for centralized corporate democracy. That's why they were so effective at managing themselves.

Finding the expertise to design these systems is a hard thing. Maybe we need to start workshops and Think Tanks to bring political science people, historians, and technicians in one room to build this kind of expertise. Iceland CAST[170], Constitutional Analysis Support Team, which was a collective of people that were interested in this approach. I don't know if they are still running.

Shermin, you also mentioned so-called smart contracts. How could we get from smart contracts to smart constitutions?

SHERMIN: Before I answer your question, it's very important to provide some historical insights on the creation of democratic governments. The democratic governments of today

170   The Constitutional Analysis Support Team (CAST), a semi-formal collective of individuals sharing an interest in the Constitution process, was established by Smári McCarthy and Eleanor Saitta in January 2011 in order to undertake analysis of the Constitution as it was drafted. The collective made itself aware to the CC, and indeed many of the Constitutional Councillors became involved in CAST's project. In particular, towards the end of the drafting process as the Constitution started to "stabilise" (mid-June), CAST arranged a Constitution "Stress Test" – an event open to citizens with a willingness to contribute to testing and finding gaps in the Constitution. The testing drew heavily on the linguistic analysis expertise of the Internet company, whose director was also a lead proponent of the exercise.

are actually unique points of failure. In the 17th century, one of the four thinkers of our modern democracy, Montesquieu[171], wrote that democracy wouldn't be possible in big societies, because there would be a political elite running the government. They would not act in the best interests of the people. This is exactly what happened.

This notion of post-democracy[172] is actually the frustration that what you vote for is not what you get. The more centralized and the bigger a government gets, and the bigger the group is that is being governed, the greater the risk is to generate a unique point of failure.

The power of the nation state is crumbling due to the effects of globalization, triggered by free trade and the Internet. I think Blockchain will be the next step of development to push this.

Political participation will be very different in the future. We are arbitrarily distinguishing between citizens and consumers. I don't know why. As citizens of democratic governments, we are consumers of government services. The only difference is that I get to vote every four years. I am not free to choose to be a citizen of another country. Yes, I can immigrate to another country by legal immigration or as a refugee. But that takes time and effort. Not everybody does it. If I've done it once, I probably wouldn't do it a second time, because it takes so much effort.

Have you heard of e-Estonia[173]? Estonia is known worldwide for digitizing all its government services. They are on the

171 *Charles-Louis de Secondat, Baron de La Brede et de Montesquieu was a French lawyer, man of letters, and political philosopher who lived during the Age of Enlightenment. He's famous for his articulation of the theory of separation of powers.*

172 *The term post-democracy was coined by Warwick University political scientist Colin Crouch in 2000. It designates states that are conducted by fully operating democratic systems, but whose application is progressively limited. A small elite is making the tough decisions.*

173 *e-Estonia is a term commonly used to describe Estonia's emergence as one of the most advanced e-societies in the world. It merges a forward-thinking government with a tech-savvy population.*

forefront of e-government solutions. They recently started a e-Residency program, which means that they are now offering citizens of the whole world to digitally apply for a citizenship, without going to Estonia. Somebody from Guatemala can apply to become an e-Resident of Estonia.

> Once nation states start to compete for taxpayers' money, I wonder how long these dictatorships will survive. <

With this e-Residency, you can incorporate a company in Estonia, which is a member of the European Union. You can start trading or paying taxes to the Estonian government. Imagine the next step, if other countries start offering such services. All of a sudden, we will be in a free market of nations. For the first time, we will have the option of democracy. Once nation states start to compete for taxpayers' money, I wonder how long these dictatorships will survive.

It's interesting to think about the notion of a citizen or a nation state itself. Bitnation[174] is a concept of a virtual nation state, based on Blockchain technology. Bitnation collaborates with Estonia and Liberland[175]. Therefore, it's still a system, dependent on a physical nation state. How is that going to change?

SHERMIN: I think it will be transitional. When we started to have streaming, with YouTube giving us the opportunity

174   BITNATION is a holacratic organization, striving to become a fully functional Decentralized Autonomous Organization (DAO). This means there are no old-school management structures or barriers to enter. Anyone can join, whether for profit or non-profit, and benefits from the infrastructure of the BITNATION community.

175   Liberland, the Free Republic of Liberland, is a micronation claiming an uninhabited parcel of disputed land on the western bank of the Danube, between Croatia and Serbia.

to put videos, online and stream them, that was not the instant death of television. Fewer and fewer people were watching traditional television, and more and more people are watching streaming formats. It will be a gradual transition rather than a radical one.

VINAY: For the very rich, we already have competition between states. If you want to incorporate your offshore tax heaven, you might go to Liechtenstein, or Panama, or maybe you go to the Channel Islands. The very rich already have choice of jurisdiction, in a practical sense.

What's left is enormous numbers of people with no assets and no ability to claim ownership of the assets that the very rich and the middle class have taken. If we want to ensure that there is some kind of actual democracy, it needs more than competition between states. It needs the citizens of the world to lay their hands upon the wealth of the rich, on the basis that it's all made from shared assets like atmosphere, ocean, and the combined knowledge base of all the previous humans.

There is no wealth that can be generated without building it from the commons. In micro jurisdictions, the elites hide their wealth from anybody else. This is the dominant form of competition between states.

I think we have to think about refreshing democracy. We have to think about it from the perspective of those that benefit from the commons. The commons should be governed by the community, which create or depend on the commons.

How could creating commons work for Blockchain?

VINAY: The bottom line is: the commons are already here. The atmospheric commons is everything we breathe. The knowledge commons is everything in the textbooks. The problem is that we don't give these commons any teeth, because we pretend that all democratic legitimacy is in the hand of the nation state.

> ## We should have improved the machinery of democracy with every new technology that came along. <

If Thomas Jefferson was alive today, and he saw us voting once every four years on paper in the way they did at the time of the founding fathers, he would beat us with a stick. We should have upgraded the machinery of democracy when we got telephones. We should have upgraded the machinery of democracy when we got television. We should have continued to incrementally improve the machinery of democracy with every new technology that came along.

Instead, we became frozen with this extreme wealth of technology. We have to do 200 years worth of updates to the core structures of democracy, all at once. Or, we must accept that democracy has become so slow and so rigid compared to everything else in society, that it simply can't provide any meaningful steering.

How do you think we can fight this inertia of updating the constitution of our democracy?

Simple law is easily amended, but we need a serious update in the constitution. How do we fight this inertia?

VINAY: There's plenty of action, it just doesn't happen inside of the nation state.

For example, the formation of the European Union is an enormous constitutional overhaul of a third of the world's economy. It's an astonishingly large and productive group of people, all participating in this fundamental constitutional change. The problem is that we aren't challenging the boundary between things.

The EU and its institutions started with one goal. The future of the world would be decided by one Europe. It was an elite project, to create an integrated Europe, so that we wouldn't have another 1000 years of war. That project is largely successful, except for the British.

> We the people
are a very disorganized,
sedated bunch right now. <

We are not seeing constitutional projects from below. We the people are a very disorganized, sedated bunch right now. I'm not sure this wasn't true in the past. I don't think that, for example, the American project was particularly representative of the people. In some ways, we are the 6% of Americans who took a side in the revolution. It was a relatively small process compared to the size of the society.

The next big push on this is to build structures above the nation-state that represent global commons, particularly climate and knowledge commons, like pharmaceutical knowledge. Then, we must beat down on the nation-states with a stick, saying, "Look, we've got 1.8 Billion people who voted to say that your country's conduct is unacceptable. Change your laws, or we'll stop buying your products. You'll go bankrupt."

We need to convene global authorities that represent the people, rather than the UN. The UN is simply a trade association for nation-states. You as an individual have no representation at the UN, except through your government. We need to build a kind of people's UN, using the Blockchain to ensure that we have one individual, one opinion. We must use statistical analysis to fill the gaps.

I would build something that was above the nation-states: a harder constitution that was modern and up to date, that encompassed new democratic voting practices. The political crisis the states are unable to solve is a global political crisis. The states are really good at keeping things running inside of their borders. But how much of the state's trouble is inside of the borders, and how much of it is in the air around us?

We're also focusing on the management of wealth, goods, and people. In the end, ideas and values move everything.

We're in a system *The Economist* calls "The Trust Machine." We once had the trust system, but we are beyond trust. Can we imprint values like freedom of speech on the Blockchain?

VINAY: You cannot have freedom of speech on a Blockchain. It comes as part of the technology package, largely because American programmers worked on much of the core Internet concepts. The American programmers assumed that freedom of speech was a fundamental attribute of the network. An American model was baked into the foundational protocols.

Think about the back-end days, when the Internet was in kind of university clusters. Each cluster had a system administrator. The system administrator was basically the sheriff of the town. They built a frontier of political economy, exactly along the lines of the one they had done on the real frontier. That is called the Internet.

> The Internet is simply the best of America, skimmed off into a layer and franchised. <

The Internet is simply the best of America's value system, skimmed off into a layer and then franchised.

Compare it, for example, to the French Minitel[176], which required a permit. It was a completely different picture of what a network should look like.

Inside of that context, if we have things like freedom of speech, we will offend everybody. 1.2 billion Muslims, perhaps, take it very, very poorly when somebody speaks ill of their leaders.

We have to accept that the Internet is going to upset people. This is what freedom looks like. People who come against that with a weapon in their hands are automatically your enemy.

Freedom of speech is worth fighting for. It's worth actively persecuting people who are trying to take it from you.

Let me add another layer to this discussion. Google says, "Don't do evil; don't be evil." Blockstream[177], on the other hand, states that it "can't be evil" because of its underlying algorithms. But who defines the underlying definition of values? Don't we need to encode human rights into the technology?

SHERMIN: Technology doesn't have a bias once you program the code. The person who programs the code has a bias and will program the bias into the code. Therefore, there will always be bias, because you have the human machine interface, and humans are corruptible.

A very good example is the Google search algorithm for faces. Google was identifying black people as apes.[178]

176 Minitel was a Videotex online service accessible through telephone lines. It's considered one of the world's most successful pre-World Wide Web online services.

177 Blockstream is a Blockchain technology company co-founded by Adam Back, Gregory Maxwell, Austin Hill, and others. It provides funding for the development of Bitcoin Core, the predominant network client software.

178 In July of 2015, it was reported that Google Photos was incorrectly labelling black people as "gorillas," resulting in public outcry.

That happened because a bunch of whites developers programmed the algorithm. They were sued, I believe.

The developers are the new lawyers in a Blockchain world. They are the definers of law and order, in a way?

> SHERMIN: They are the definers, yes. If you want to become a lawyer in the future, coding will be part of your education. We are entering a world of automated smart contracts, and the basis of this will be codes. This is already happening. Within the Bitcoin network, the Bitcoin protocol is already a kind of automated legal framework. Algorithms are already ruling our world and are part of our economy.

We are talking about a decentralized system. Is it supporting or improving democracy in a way. How does this work?

Today, when I want to read the law, I can open the constitution and read it, whether I'm a lawyer or not. If I cannot code, then I'm not able to contribute or understand what's written.

> SHERMIN: Why aren't we teaching code at the universities? At the BlockchainHub, we're stating that we should foster to learn how to code as early as possible. We should start learning how to code in kindergarten. It should have a fair share in our educational system. It's the 21st century, everything is running on code, and we are still not teaching it on a basic level. There's a gap between what is taught to society and what we need. Just imagine a world where we only learn Latin or Old Greek. Today we need to learn how to code.

>We should start learning how to code in kindergarten.<

At the same time, you have to consider that the history we built upon is based on Aristotle. If you lose the ability to read ancient Greek and understand Aristotle, you cease to understand or recreate the foundations of civilization.

VINAY: That's absolutely true.

It's not a competition of education; it's a development of education. In Sweden, Israel and a couple of other countries, people are coding from the first class, on. Hopefully, we'll see that in Germany, too.

It is a competition of ideas, of values. How do we define that in democracy in this new era? Democracy is a process of creating or gathering the opinion of many to execute it for all. How will Blockchain do that?

SHERMIN: If we fast-forward, the distinction between citizen and consumer will dissolve.

We might be citizens of different Blockchains or users, and participants of different Blockchains, rather than citizens of a certain nation-state. The whole notion of the nation-state is already obsolete. In probably 50 years, we won't see it the way it is now.

You've referred to a value system, and I would like to refer also to our economic system; it is going to change considerably.

Read Jeremy Rifkin's *Zero Marginal Cost Society*[179]. The way we create value and generate value out of economic production is not going to exist in its current form and will fundamentally change in the near future. We are entering the age of zero marginal cost. In the crypto world, you can create your own dollar, and you can create your own crypto currency. You can define the tokens of your crypto world. The tokens can be defined as a new set of values.

Let's say I create value by being ecologically conscious. There's a Berlin-based start-up called Changers, and

179   *In the book The Zero Marginal Cost Society, author Jeremy Rifkin describes how the Internet of Things is speeding us to an era of nearly free goods and services. In the book, this causes the rise of a Global Commons and the eclipse of capitalism.*

they're not just on the Blockchain. They have an app that tracks how much CO2 emission you save by going by public transport rather than by cars or planes. You get tokens for saving CO2 emissions.

The app creates a set of values, which is a model that can be interchangeable for other things. We have to reset our brains and think in new value systems. The system we have right now is arbitrary. We think it's the only way, but there are so many other ways.

When the Internet started, we had the notion that the technology would be decentralized, would improve democracy.

Today, we have a massive advertising system, which is also a perfect surveillance system. Besides it created incredible monopolies. How do we make sure that a Blockchain technology won't have the same destiny?

VINAY: The underlying problem is that, to be honest, the nation-state hasn't really existed since the end of World War II. In actual fact, there are only six actual countries in the world. They are defined by nuclear weapon stockpiles: the Americans, the Israelis, the Chinese, the Russians, these large, transnational blocks, the French. These blocks define actual sovereignty.

> ## >Countries or nation-states are just fictions.<

No nation without a nuclear weapon or a defence pact with another country with a nuclear weapon is actually sovereign. They are regions; they are just fiction. If you throw them away and simply work on nuclear umbrellas, defaming the zones of geopolitical influence, the international system goes from being extremely complex to completely transparent.

South America and Africa are complete messes, because the nuclear powers take what they like from them. There's nothing the local governments can do. They are not sovereign; they have no actual authority. To expect Blockchains to fix the political mess left by nuclear powers is unrealistic.

The NSA is a thin mask over the American nuclear state. The American nuclear state has operated inside of the American democracy for 60 years. It says, "Thank you very much for the tax budget," and then defames the fundamental terms of engagement in the society. There is no global movement to get rid of the nuclear weapons with any degree of weight. There hasn't been since probably the late 1980s.

There's no way we can fix the world, divided into nuclear fortresses, with a bunch of software written by a team of maybe 200 people. It's just not realistic. We might put pressure on the mayor; we might persuade some small countries to bear our trade or immigration policies.

Nothing will change until you see a global will among the people, saying, "Look, running the planet so that we all survive is more important than having these zones of geopolitical influence backed up by our local nuclear arsenal." It's almost like the Berlin Wall never fell, the Cold War never actually ended. We are still in a world of geopolitical conflict, backed up by nukes.

We have the illusion that it changed because of the American economic boom of the 1990s. There's a single line of transmission from Google's advertising model through the NSA back to the American nuclear state. The Internet is just an enormous eye for the NSA. It's part of the security apparatus.

We won't be able to take these tools off people and use them to beat them at their own game. It's just not the way this goes. Tor Project, with all of its freedom advocates, is funded with roughly two million dollars a year by the American government. Are they doing that because it's a threat to their power? Of course not.

We need some hard, political realism about how the world runs. A billion people are starving; there's a massive concentration of wealth and privilege. The wealthy won't permit political change, because it would make them radically poorer.

Some interesting software is not fundamentally capable of fixing this. If there was a mass, political will to produce real fundamental change, to ensure that every human being is fed, has basic healthcare, has a right to an education, and has the right to contraception, the Blockchain could certainly help.

Bitcoin received value because the American libertarians brought political will. The technology then received value. Unless you see very large-scale movements all across the world to do something about the deep ecology and the balance of power between rich and poor, it doesn't matter what technology you introduce. It will be used to shut up the existing system.

We have the ability to feed everybody right now. We've produced twice the amount of food that it would take to feed the entire planet, just today. People ignore that information.

Shermin, as a Blockchain advocate, will you talk about the disruptive potential and how that political will and power may shift from the powers that be to the powers of civil society?

SHERMIN: Blockchain is a game changer, just like the Internet was. We will have the opportunity to remodel our society through cooperative, decentralized, autonomous organizations that have new sets of rules. I'm very excited to see what this collective intelligence of eight billion people will produce.

I have to confirm what you're saying: people are people. They are, very often, five-year-old children with big egos, who fight each other. We also see this in the Bitcoin community.

Blockchain is a means to an end. It's a tool, and we can do great things with it. There's always the human component, however. It's not the answer to all questions, *The Hitchhiker's Guide to the Galaxy*'s 42.

It's a cool tool, but we still have to work a lot. We have to learn by doing. We will learn with every new Blockchain. We will learn from errors. We need patience.

There's one monopoly that we gave to the democratic system. It's a monopoly of life and death, of violence. How would that work in a Blockchain society?

VINAY: Blockchain could support almost any social system. I've done the analysis for capitalist anarchy. I've done the analysis for socialism. I've done the analysis for democracy.

Over the past 20 years, there have been many discussions of assassination politics. It suggests that you basically crowdfund executions. I don't really see that as a credible thing.

The American model originally operated with the idea that you privately fund wars by funding collective defence. This is not so different from the Swiss model. This intersects other fundamental issues, because, at the end of the day, until your Blockchain government is paying the salaries of people who apply physical violence, you are not going to have control of those people.
If you have a democratic mandate, and you use the Blockchain as a way of implementing your democracy, nothing changes. You should build different democratic forms using a Blockchain or even post-democratic anarchist forms. It's worth the experiment to use it to

implement governance for a merge-base, which is coming in probably 10 years.

Think about the algorithms we actually use today and the high frequency trading at the stock market. The algorithms are capable of ruining the stock market in just a few minutes. Half a year after the ruin, we have problems figuring out what happened.

Let's imagine some sort of DAO led by a very crude AI, making smart contracts with other DAOs that no human lawyer can read, because they can't code. They make decisions that are capable of ruining the world, without a care. What's your opinion on that?

> VINAY: That seems a pretty fair description of how it would look if it went wrong, yes. Charles Stross'[180] science fiction novel called *Accelerando*[181] is about a scenario very much like this. It's really good; it's quite scary.
>
> SHERMIN: It could go that way. I'm not a cryptographer. I think these are questions to ask AI people. This is actually the worst-case scenario, the Skynet scenario you just described. This is why we need to discuss this on a broader level.
>
> I believe in the people who design these core protocols, but we need to get more people involved from all disciplines of society.

When transactions on Blockchain are open, and I put a contract on it, it will be open-source. Nobody can have ownership for this intellectual property, because you can read it openly. Everybody can write on it. The decentralized network will verify it.

180   *Charles Stross is an award-winning British writer of science fiction, Lovecraftian horror and fantasy.*

181   *Accelerando is a 2005 science fiction novel by Charles Stross. In the book, the title refers to the accelerating rate at which humanity in general, and/or the novel's characters, head toward the technological singularity.*

Can you predict a Blockchain that isn't open and not commons, regarding the content on the Blockchain? What are future business models in a decentralized peer-to-peer net?

SHERMIN: The whole idea of the Blockchain is to be open source. There's the consortium Blockchains, where you can have things more private. The public Blockchain only works because it's open source, otherwise the network can't verify transactions.

Look at the use of Airbnb. This is open source, what Airbnb is doing, because they can't patent their transactions. People who use the Airbnb platform, either as a renter or a landlord, could just copy these transaction models, put them into code, and put them on the Blockchain. Airbnb as an intermediary would be obsolete. With a very good user experience and some marketing, Airbnb would be gone.

This is actually discussed in the book, *Zero Marginal Cost Society*. It's already happening, independent of the Blockchain. This proprietary thinking, with closed systems, is out-dated.

> ## >I just don't think there's political will for radical change.<

You present Blockchain technology as a tool that will change the world.

What is the actual strategy or process that you see this happening with? The world is not only technology. It plays a role, but it's just one thing amongst many other things. How you see this revolution happening?

VINAY: I just don't think there's the political will for radical change. I don't see it in the populations. They will complain, but they will not act. I don't know when that will change.

That doesn't mean that it will not bring transformation. There's much thinking about this eco-utopian vision and how we get there. Nobody is quite willing to start tearing up the pavements. It all depends on the personal, political will of each person.

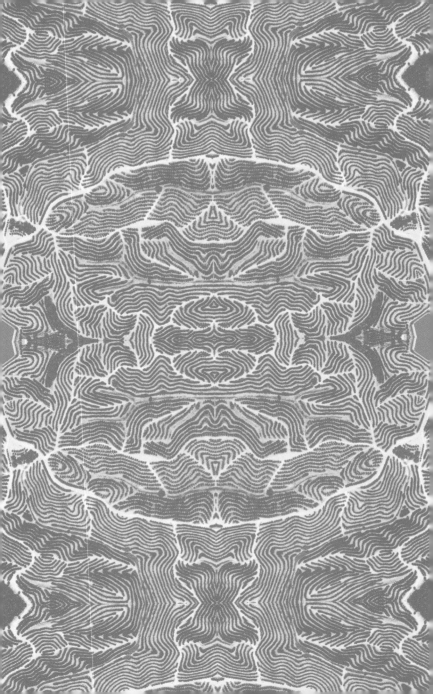

7
The Brave
New Art World

Masha McConaghy
Elizabeth Markevitch

PANEL DISCUSSION, BERLIN, JULY 7, 2016

With the shift toward digital media in the creative world, new forms of creative expression arise. How does the experience of an original or virtual artwork translate into the digital space? What defines a digital original and how does it find its audience? Who will be the driving and defining forces behind this paradigm shift?

Everything on the Internet is free. Therefore, how can creators secure their intellectual property and benefit from their talent? On the other hand, how can art lovers support artists of their choosing, inside or outside of gallery circles? Until recently, owning a digital artwork was a challenge, as the art world struggles to adopt new technologies. Digital exhibitions and streams have yet to reach mainstream society. But as the traditional gatekeepers of the art world change and lose their monopolies on art interpretation and distribution, new protagonists arise to create a brave new art world.

Blockchain has already challenged the financial world with the appearance of crypto currencies, like Bitcoin and Dogecoin. It will further enable new forms of registries, certificates and value as well as property management. We have yet to see how these developments will further shake the foundations of the art world.

On the D.DAY panel, Blockchain expert and ascribe.io co-founder Masha McConaghy and art professional and ikono.tv founder Elizabeth Markevitch discuss these monumental shifts in the art world. The pair collaborates on exploring the boundaries of the art world. Their contributions have already altered the very core of art reception.

PANELLISTS

MASHA MCCONAGHY, curator and researcher, is a co-founder of BigchainDB, a scalable Blockchain database, and ascribe.io, a service enabling immutable attribution for artists, offering clear provenance for digital and physical art. She has a PhD in Arts from Pantheon-Sorbonne University, Paris and a Museology Degree from the Louvre School, Paris. Her PhD explored the distinct relationship between art and commerce over the centuries. She has organized exhibitions throughout the world, worked with curators at the Louvre Museum, Paris and directed a commercial gallery in Vancouver. Her current pursuits are a cross between art, IP and applications of new technology.

ELIZABETH MARKEVITCH is an art professional and the director and founder of IkonoTV, an international platform for broadcasting visual arts. Markevitch began her career in the early '80s as assistant fashion editor for Vogue Hommes and has since served many roles in the art industry, including: director of the Art Fund, Artemis; director and founder of the art advisory department of J. Henry

Schröder Bank; and senior manager of the painting department at Sotheby's. Markevitch works as an art consultant and has collaborated and curated a wide variety of special art events. She is a pioneer in the digitization of the art world and is currently launching ikonospace, a revolutionary new 3D software for gallery curators, art fairs, collectors and museum exhibition designers.

Masha, you co-founded the company BigchainDB, which is a company providing Blockchain database solutions like ascribe.io. How did you come up with the idea of the company?

MASHA: My quick explanation about Blockchain is basically the following: think about Blockchain as a spreadsheet in the sky that no one owns, no one controls, you can only write on it, you can never delete from it, and the records are there forever.

Let's not get into the discussion of what forever means.

The initial application on the Blockchain was Bitcoin, the first crypto currency based on Blockchain. This was revolutionary because all the attempts before digital e-money fell apart. All the transactions were bound to a centralized database. When the company went broke or their records were tampered with, people lost their money. But Blockchain has no centralized control over all spreadsheet.

> Think about Blockchain as a spreadsheet in the sky that no one owns, no one controls. You can only write on it, you can never delete from it, and the records are there forever. <

And what problem does the Blockchain-based application ascribe.io solve for the art world?

MASHA: About three years ago, the question we asked was: how do you collect digital art?

My partner and I travel a lot. We know many artists, and a lot of these artists work with digital tools. We asked ourselves: how can you own a digital art piece? Back in the days when I ran a gallery in Canada, the most digital collectors would ever go was digital print. When I asked collectors, "Why wouldn't you collect video?" the reaction was, "I don't know how. Is it just a cassette? What exactly am I owning?"

My partner and I brainstormed about it. Trent was very involved in the tech world, following Bitcoin right from the start. He was fascinated about technology. He came up with the idea to own digital art the way you own Bitcoin. It seemed like a very good match.

Elizabeth, you come from a very traditional art background. In the first positions you had, the Sotheby's and the bank, you were dealing with traditional mechanisms. What made you shift to ikonoTV? What was the reasoning behind starting ikonoTV?

ELIZABETH: In '97, I had another idea, and I was obsessed. Up to '97, I was an expert in Impressionist art and modern art and working with these big companies. I was in this very elite world. I was asking myself how I could share my passion for the arts.

Because I was still very market-orientated, I came up with the idea on how average-income people could buy a work of art from a famous artist, without necessarily having the money to afford the original artwork. With two friends, I built EYESTORM[182]. EYESTORM.com was the first online gallery, built in '99. Today, we are a dinosaur in the digital age.

The concept would not have been possible without digital media. We asked artists, very famous artists, to give us a digital file of their work and to reproduce it on photography paper, so people could buy it online and own something nice for a more affordable price.

182   Eyestorm is the leading online retailer and publisher of limited edition contemporary art, offering work by both established and emerging artists.

Funnily enough, even though the gallery was on the Internet, only people from London and New York were buying.

From there, it took me years to develop the idea of ikonoTV.

One day, I had the idea to compare the art world systematically with the music industry. In this comparison, I understood that EYESTORM was basically the producer of the record. It still produced a physical object.

A record is a product, and this product is worth the same, independent of what the famous artist is doing or not. The artist has no control about distribution, who is buying it, or who their fans are.

Staying in the metaphor of the music industry, the art piece was as if you could only experience music today in a public space, in a bar, or on the radio.

IkonoTV is basically the radio. It's trying to bring art to everybody's home to bring people into art world. It's a first step for people who have no clue about art, who are maybe not even interested in art because they think this world is only for real art lovers.

The main idea behind ikonoTV is to provide direct access into the art world, since the art world has this tendency to intellectualize everything. It can be overwhelming for some.

The radio had a lot of resistance from the music industry in its early days. I imagine you had a lot of resistance. Where was that coming from? From the artists themselves or the surrounding? What were the first challenges?

ELIZABETH: Nobody really understood what I was trying to do. I started EYESTORM in '99 and ikonoTV in 2005. It took me five or six years to knock on doors of artists or the doors of museums. I'm still doing it. It's my main job.

Already, at that time, the word digital caused irritations. When I started with ikonoTV in 2005, it was the same year YouTube came out. We have the feeling YouTube has existed forever, but it's only since 2005. That was exactly the same year I started.

Can you explain in two sentences what ikonoTV does for all of those who are not familiar with the concept or the program?

ELIZABETH: I would like to take another parallel to the music world to make it more tangible. We basically provide a video stream, which runs continuously. Some people call ikonoTV the MTV of the arts, but it's more in the spirit of MTV in the eighties, not MTV of today.

MTV, when it started, was a real revolution for me. I was pregnant with my first child, and it was overwhelming, because, at that time, we only had three public TV stations. Then, suddenly, MTV arrived. It was like a shock.

MTV did something fantastic. It invented what we call a video clip today. It's the marriage between a musician and a filmmaker, creating a third form of art. We do exactly the same today at ikonoTV. I have art historians, artists, curators, and protagonists of the world of art. I ask them to tell the story of the art piece to the filmmaker, not to you and me.

The filmmaker then tries to translate this story into a dynamic visual form. What you see today is moving images on a single work, but very slow. We give you time. With this is the possibility to contemplate the artwork and rediscover a specific detail, the technique. We tell you the story about the art piece visually.

In the beginning, you knocked on many doors, talking to artists to persuade them to give you files on their art pieces to use them for ikonoTV. This required a lot of convincing and building trust.

This was the main reason why ikonoTV interlinked with ascribe.io, right?

ELIZABETH: Yes, it is a lot about trust. It's not necessarily the trust in how good we would do our job, but trust in digital dynamics. It's the digital world, which make people very afraid. It never occurred to me as a problem. When you use only low-resolution images, what would people do with them?

You have the same problems with analog mediums. 15 years ago, an artist told me that a guy from Japan came to his studio holding a new book about his work, which included photos. He didn't ask him for permission before making the book. Basically, he stole the images. Some images were even scanned from other books.

Therefore, nobody mentions the trust issue when we talk about analog media. We do it when we talk about the digital media. And as soon as you say the word "digital," everybody says, "No way. I want to protect my works."

MASHA: It kind of scares people, because they have this sense of losing control as soon as you say the word "digital" or "Internet," because people have this notion that everything you put on the Internet goes away. And, that you can't really catch it anymore. With this, you risk to blur the link between the file and the creator. Just imagine how many images we have on the Internet that we have no idea who created them.

I think Elizabeth is an absolute pioneer in the art world. She's probably the first person I met in that context, who, when discussing Blockchain technology, immediately understood how ascribe.io would make sense as a solution for artists.

> Because digital media blurs the link between the file and the creator. <

ELIZABETH: It makes total sense to provide security to the artist about the usage of their digital files. IkonoTV doesn't need ascribe.io to run our service, but the artists or the museums need ascribe.io to trust the digital world. I encourage them to use ascribe.io. This makes my life more complicated, meaning I not only have to explain what I do at ikonoTV, but also what ascribe.io is doing.

It's really important to make them feel comfortable. You tell them to sign up at ascribe.io, get an account, upload your digital files, and define the rights how others can use your work. You simply give me the right to stream it.

So ascribe.io is a platform that tries to manage ownership about an artwork in a more substantial way. So it's about provenance?

What can you actually do on ascribe.io, Masha?

MASHA: The Blockchain technology is the central technology at ascribe.io. Blockchain is perfect for tracing provenance, because you have this record about where the file came from and where it will go. It is a very clean ownership lineage, or ownership provenance.

You could say that Blockchain is like a time stamping mechanism. In this sense, an author can time stamp the claim of authorship. Ideally, you do this at the point of creation, for proof.

But it's not just time stamping. At ascribe.io, we also built standard licensing methods and standard contracts that allow you to transfer certain usage rights to another person. For us, this was important. In a digital world, on one hand, you might be scared to put your files on the Internet. But on the other hand, a lot of artists want to spread their work online. At the same time, they want to monetize their work. This is this difference between title and copies, or access and title. Blockchain helps you secure the title.

> In an analog world,
art is valuable because
nobody can access it.
In a digital world, artists price
their work based on how
many views it gets. <

Let's say you are the creator of video art. On the Blockchain, you time stamp your claim of authorship. You create five unique editions, and then you transfer one edition to ikonoTV for streaming. At the same time, you're a very open artist. You want to spread your video around the world, because that's what is kind of the new proxy for value. The more people see it, the more value it gets. This whole digital shift also influences the process of value creation in the stress field between scarcity and abundance. In an analog world, it's valuable because nobody sees or can access the work, but in a digital world, some artists price their work based on how many views they have.

You're creating a mechanism to manage and distribute art. Since you are a former manager of a gallery in Vancouver, do you think ascribe.io can become some sort of online gallery system itself? Can it enable a digital art market?

MASHA: We never wanted to be the front end, to be visible. BigchainDB and ascribe.io are focused on the technology side. We're more the plumbing in the background. Like PayPal is a payment processor by email, ascribe.io is a digital rights processor by Blockchain. We want to empower the front end, the user experience provided by platforms like ikonoTV.

It could become a marketplace if somebody wanted to build a front end.

MASHA: People do want that. For example, we had one 19-year-old who contacted us by email saying, "I'm going to do a marketplace for digital goods, and I need an ownership processor of the rights."

I asked, "How long you are going to take to build it? We talked to another place that's been building a marketplace for a year already."

He said, "Okay, I'll use Shopify[183] as a front end. Then, I'll just integrate you on the backend. So I'll be done in two weeks."

We're like, "What?"

He said, "Yeah, the marketplace will be done in two weeks."

This is what is so amazing with digital tools. They allow you to experiment with things to see what works, what doesn't work, etc. We are empowering other marketplaces.

Recently, another online art marketplace came up with the Lumen Prize for Digital Art[184]. The Lumen Prize for Digital Art from U.K. became an important art prize for digital arts specifically. They strongly support their artists, saying things like, "We receive so many wonderful artist applications, and we want to help all of them to monetize somehow." But how do you do that? They just opened a new marketplace, where you can buy digital editions from their artists.

A lot of people want to help artists monetize, and you just need different tools and different places to put them together and build something wonderful.

183   Shopify is a Canadian e-commerce company that develops computer software for online stores and retail point-of-sale systems.

184   The Lumen Prize is an international award and tour for digital art. Its goal is to focus the world's attention on this genre of fine art through an annual juried competition and global tour.

Things are changing fast. It's exciting.

Elizabeth, when you talk to museums, when you go to the art fairs, when you talk to galleries, what is your experience in the willingness to change and, at the same time, the need to change? What sort of ecosystem do you see in the future? What role will you possibly play?

ELIZABETH: Museums are still very slow. Specifically European museums are the slowest. They think they're super modern by just having a website. So they have a website, but they still haven't digitized the collection. In America, what I'm doing seems so natural. When I talk to them, they say, "Yeah, we see immediately what you want."

> European museums think they're super modern by just having websites. <

Since Masha just mentioned online marketplaces, ikonoTV just launched another marketplace three weeks ago. It is our On-Demand section, where we established the Freemium[185] model. You have free content on-demand, but also premium content, which requires a monthly subscription fee.

Here, instead of selling art, we will pay royalties and licensing to the artist, because we will always be streaming. Through us, you will never own a work of art. The artists sign a contract with us on ascribe.io. They agree that, in fact, we can stream, and they can limit it for a specific time: one year, two years, whatever. Then we stream it, and they have a revenue share from these subscription fees.

185   Freemium is a pricing strategy by which a product or service is provided free of charge, but money is charged for proprietary features, functionality, or virtual goods.

It's interesting that you call it a marketplace, because you could also call it a virtual gallery. How will the notion of a gallery change through this new virtuality?

> ELIZABETH: The gallery system is changing on its own. The gallery system will slowly disappear; it doesn't need digital medium to do so.
>
> I don't think change depends on the digital medium. I think it depends on the artist's needs. Artists need bigger spaces. They don't want to always exhibit in the same space.
>
> As a gallery, you ask an artist to fit his or her work into your space. You give them a restriction, a framework that is already limited. But as a gallerist, you should be able to do the opposite. You should say, "What kind of space do you need for that work?" And you should provide that space.
>
> The problem is: we know that every seven years, as an average, a gallery should move. After two exhibitions, the artist is fed up. Galleries lose their artists. They go to other galleries with bigger space or whatever. Then, they also leave that bigger space.
>
> Change happens when there is frustration in the air for the artist. This is why I think the galleries themselves have to rethink the way they work.

We also see a form of liberation in the expression of an artist through new technologies. How do you see that?

Will we experience new identities of the artist and how the artist adapts to the gallery or museum space, or through limitless space in a virtual arena?

What do you think is happening to the identity of the artist and how he expresses him or herself?

ELIZABETH: I don't think the basic characteristics of artists will change.

When we talk about the digital medium, it's all about ownership. But do we need to own a work to look at it? That's the question. We are in a world where we're always talking about selling and buying. But what I'm doing is not about selling and buying; it's about access to art. With different tools like 3D, we are providing new types of experiences. These new types of experiences will never kill the real experience. It's not either/or.

Using the music industry as a metaphor, you have four types of experiences. Let's say I want to throw a big party, and I ask Daniel Barenboim[186] to come and play on my grand piano. For this experience, I must pay many thousands of Euros to have him at my home. Only 15 or 20 people will enjoy an amazing, private, live evening, with an amazing musician.

You could also say that this is ridiculous. "You live in Berlin. Why don't you just go to the Berlin State Opera? Barenboim is there every day." Seeing him at the opera would be another form, experiencing him with many others. Or, I could go and buy his CD and own the CD. A fourth way would be to just turn on the radio or Spotify. It's not either/or. These experiences can exist in parallel to each other. I think in the art world it's the same. It is about experience first.

We often refer to Marcel Duchamp[187]. He described this wonderfully in his manifesto in the early '50s. He said that there is no work of art if there is no viewer or no spectator to give the work social value. And that is true. If you're alone in your bathroom with your work, nobody will see it. Meaning your work doesn't exist. In this sense, digital

186   Daniel Barenboim is an Argentine-Israeli pianist and conductor who is also a citizen of Palestine and Spain.

187   Marcel Duchamp was a French-American painter, sculptor, chess player and writer whose work is associated with Cubism, conceptual art and Dada.

media only provides you another access to reach out to your spectator. Art lives complementary. Digital media make the artwork live more. The more it is seen, the more it lives.

When we talk about art, there's always the notion of scarcity, access, being able to own or see that artwork in a limited circle. That was part of the experience. That may be changing. Even eBay and Amazon are creating art platforms. And even the Gagosian Gallery has a "click and buy" button. Is this a main-streaming of the art industry?

ELIZABETH: They're not inventing anything new here, I'm sorry. They're just using technology to allow you to buy something quicker.

What does it mean for the arts? What does it say about the quality of the arts if, let's say, art is defined by review?

At the publication Flash Art[188] it was always about who said what about which work, which also co-defined the worth of the work. What people said ultimately decided which collection it would enter and who would buy it. For example, the biennials[189] et cetera, all contributed to the worth and value creation of art.

How well can ascribe.io accommodate these reviews these value processes?

MASHA: Coming back to the galleries, I actually don't think they will disappear. I think their work field is shifting and with this, their role is shifting. There is always this human aspect bound to institutions. Some collectors like to belong to one club of people, and others to another. Or people prefer a certain gallery experience or event experiences.

188 Flash Art is a magazine featuring artwork.

189 Biennale is any event that happens every two years. It's most commonly used within the art world to describe large-scale international contemporary art exhibitions.

Still, the most important element is actually the curatorial aspect, especially when we have so much content. There is even more need for curation. The curator as a profession became a modern phenomenon of the mid-20th century. The exhibition Magiciens de la terre[190] in 1989 was the first exhibition that showcased this.

Galleries still have a big role in curation, because a lot of people don't know what to get, so they ask the gallery. It's like you're going to different types of stores with different types of clothing.

Galleries also have an educational role towards the collectors and in developing the career of the artist, helping them grow.

ELIZABETH: Everything is curated. It is like radio. You curate your playlist. This is part of the process of creating value as the curators, as the gatekeepers to exhibitions, to galleries, to museums, to biennials, and to critics. They are the voices that validate art. How will that play out in the digital world?

MASHA: When you use technologies like Blockchain in the regular art market, authenticity and chain of ownership can be proven, which brings the value of the artwork up.

Let's say there is a Van Gogh. One expert says it's a Van Gogh, and the other expert says it's not. What Blockchain technology enables is a time stamp of all those claims, together in one place. Every time one expert says or gives a report about the artwork, it will be attached to the artwork, which provides you this whole archive of all the different experts. What happens in the traditional art world is: one expert stated something, but it was lost. It needs to be re-expertized. This archive of

190   Magiciens de la Terre was a contemporary art exhibit at the Centre Georges Pompidou and the Grande Halle at the Parc de la Villette in 1989.

knowledge of valuable proof can be transferred with the ownership for a clean provenance.

Blockchain is just a mechanism that helps to address a certain problem in the art world. Blockchain itself is not going to prove that an artwork is real, but it can certify it by time stamping the claims.

ELIZABETH: I'll give you another practical example. One of the best artists who did a fantastic job to organize his works was Paul Klee. Paul Klee was maybe the only artist who put a number on each artwork. Through this, he built up his Catalogue Raisonné. Everything we do today to archive an artwork is something he already did.

That's what ascribe.io is doing. It gives you the number and information. It makes the artist's life easier. Once it's ascribed, it's done. How many times have we asked an artist to provide us all the relevant information, and it's lost? They always have to repeat the process to get all the relevant information together, which means that they repeat the errors.

If they do it only once on ascribe.io, they have a trace of everything they have done. The meta data of the work is done only once. The artist can correct it, of course. They can change it, if they want to. It's completely under their control.

I signed up on ascribe.io and set up an account. I uploaded a random picture from my computer to test the service, but I, of course, didn't create the picture. Now, I'm certified on the ascribe.io Blockchain as the creator of this picture, which is false.

In the terms and condition of ascribe.io, it states that things on the Blockchain are not deletable. How do you prove that the person who uploads the digital art piece is really the rights holder?

MASHA: We don't check if you're the creator. We say you have to be the rights holder to register. If you're not, you're committing a fraud.

>But technology is not replacing the ecosystem. It just offers more evidence.<

That's why it is important that people register their work at the point of creation. Then, you have the proof that you had access to that file first. Some people will also register on behalf of an artist, with the artist's name. But if it goes to the court of law, and if you committed a fraud, we can do tests. We can actually delete the possessorship, but even the deletion will be in the Blockchain.

If you want to transfer the rights to someone else, they say you are not the rights holder, and they can prove it on the Blockchain, then they can take you to court.

But technology is not replacing the ecosystem. The courts are still working. It just brings more evidence. When I explained ascribe.io to several lawyers, especially copyright lawyers, they think it's a fantastic tool for them.

Let's look at the production side of art. Take, for instance, Andy Warhol. He took existing designs from mass media and advertising and reproduced them to create his own expression of pop art. With this in mind, could we also put the creative input from *Campbell Soup*[191] on the Blockchain? Someone could say, "You took my original and reproduced it, but I don't get any money out of your sales."

With all these forms of artistic collaboration, shouldn't there be new ways of tracing the creative input of a final art piece?

191   *Campbell's Soup Cans is a work of art produced in 1962 by Andy Warhol. It consists of 32 canvases, each measuring 20 inches, each with an image of the canned soup the company offered at the time.*

ELIZABETH: If you're talking about Warhol and his wallpaper, he wanted everybody to have his wallpaper at the same price of wallpaper. But today, you can't afford one square meter of his wallpaper. I'm sure he wouldn't be very happy about that.

At the end of the day, it's what Masha said. It's not technology that makes a process more honest. There is always a human factor.

The question remains about what the digital original actually is. In the end, we have digital ledgers to re-create and multiply ownership. To the general public, the artwork looks the same, the file is the same, and the printout is the same.

How do I secure a digital original?

MASHA: We differentiate between two statuses. On one hand, there's the original *title*, and on the other hand, there's the *copy*. After certifying the original title of a video art piece with five editions on the Blockchain, it's all about transferring rights, not about the copies.

If I transfer the rights to you, I transfer the rights of title to you. In this case, you have a copy. Feeling like you're losing control because you don't know where the copy is becomes invalid. With the Blockchain, you can secure the provenance of title.

We created another tool called WhereOnThe.Net, which is a search engine to find your copies. It works with photos or images. You can paste the image URL, and it shows you how the copies are spread throughout the Internet.

It's like a search engine, but it includes analytics. We did a couple interesting examples with logos of companies. The companies can see exactly when they put their logo online and how the image started to spread. Now, they have like 8,000 images. Then, you can dive deeper and check the URLs of those images. This helps you have an idea of where the copies are.

> ## > Artists want their image spread, but they want to know who is using their art, where, and how. <

It gives you a sense of control. A lot of artists want their image spread all over the Internet, but they would like to know who is using their art, where, and how.

When we talk about new technologies, like virtual reality, the space seems endless. Normally, when we go into a gallery, or a museum, art pieces are framed. Within virtual reality, there's no longer a frame. The art piece has unlimited space to expand.

Will we see new forms of art emerging with new forms of technology?

ELIZABETH: That's for sure. But what this will be exactly, I don't know. That is the nature of an artist, to take a space, if it's virtual or not, and do whatever he or she wants to do with it.

Today, digital media also provides public information— but it can't provide everything. I loved the campaign that Tate[192] came up with, which said that museums give you answers that Google can't. And it's true; you still have questions that Google can't answer. You can only get that in the real world by talking to a real curator, talking to a real artist, going to an exhibition, or listening to a talk.

192   *Tate is an executive non-departmental public body and an exempt charity. Its mission is to increase the public's enjoyment and understanding of British art from the 16th century to the present.*

> That's the nature of the artist.
To take a space, virtual or not,
and do whatever he or she
wants with it. <

Why are talks like this so popular today? Because people still want the real thing. Why do we still go to yoga class, if you find online yoga courses on YouTube for free? It's never the same. There are things that a teacher will tell me I'm doing wrong that I couldn't learn with the online course.

I remember a conversation we had about your new venture, ikonospaces, and the power it gives to curation and exhibition. It becomes a time machine.

Will you discuss that, the new dimension, and how it offers new possibilities of experiencing art?

ELIZABETH: It's complementary. My son came up with software that was originally for real estate. It helped to either build, buy, or redo an apartment. You see a space, for instance, on the market, but you're not in it. You don't know if your kitchen table will fit in that kitchen. With the software, you put on your VR glasses, and you are in the room.

When my son showed me that with a mouse click, I could change the painting on the wall, I said, "Stop! What you've just done is the perfect curator's tool."

As a curator of an exhibition, you face the same problem. You have ideas in your head, and you know exactly how you want to do the exhibition. But how can you best describe it to the people who will be hanging it?

Then, you try to do it with a 3D model, but you need the architect who uses the AutoCAD software. Still, you don't know if your exhibition really is perfect. You only know when you're going to hang it. You have to face it.

We call our solution ikonospace. With ikonospace, you use the images on your computer and put them on the wall with all the details regarding the measurements. Then, you pass the details on to the hangers, either in 2D or 3D.

We even proposed new exhibition concepts to museums by telling them to rebuild two temporary exhibition rooms. You put 100 works in it, but then you ask the public to play with it and do what they think should be done with these 100 works. With the software, you can now interact with the public.

>The art exhibition software can now interact with the public.<

At a conference last week, I heard Thomas Girst, the Head of the Art Department for BMW, talk to a crowd of museum directors. He said, "You should have a room in your exhibition for the public to curate and interact with the art works." But who would have the courage to do that? Maybe it won't necessarily happen in real life. Why not at least virtually?

If the virtual space is unlimited, you can do whatever you want. You can see previous exhibitions that everyone uses as a reference. With this tool, everybody can experience it. It becomes an archive for the museums.

What direction do you see the art world going with the digital shift? What is the most exciting idea that comes to mind?

MASHA: I really like technology, and I'm very excited about it. But I think that technology is a tool. Artists are artists, they are the creative minds; they just use different tools.

It's hard to say how artists would evolve. The tools will become much more advanced in terms of technology. I think, as a next step, we need to get rid of this digital shell in the art world. Art is just art; it's just going to be done with different, digital tools.

ELIZABETH: Honestly, in the time we're living today, in terms of terrorism, politics, fascism, we are all lost, regarding values and purpose.

In the museum world, we must take a role, because artists are the best ones to reflect on what is going on around us. Politics know that. That's the reason they push them away and don't want the artist to interfere too much. Museums definitely need to take on this role today and stop being just mausoleums.

Tate is also a great example for audience experimentation. When they did the Sun exhibition from Olafur Eliasson, the people took over the space. It was theirs; it was not in the control of the museum directors and curators anymore.

Curators are here to create a direct dialog. We will probably see a shift towards performing arts or mixing arts, between music or fashion or theater.

> Artists are the best people to reflect the status quo. Museums need to take on this role and stop being mausoleums. <

THANKS!

Alexa Clay
Allison Krupp
Andreas Gebhard
Angela Richter
Anne Jost
Ariane Conrad
Asmaa Guedira
Bianca Preatorius
Boris Moshkovits
Caroline von Reden
Chris Glas
Cristina Riesen
Damian Gerbaulet
Daniel Reiter
David Sonntag
Deanna Zandt
Diego Calvo
Dominik Kenzler
Elisabeth Stangl
Elizabeth Markevitch
Geraldine de Bastion
Hanna Hünninger
Ilonka Moshkovits
Ivonne Dippmann
Ivonne Greulich

Izogie Guobadia
Jacob Appelbaum
Jens Pieper
Jonathan Imme
Joseph Huff-Hannon
Judith Orland
Manouchehr Shamsrizi
Marek Tuszynski
Masha McConaghy
Matt Greenberg
Monika Frech
Myriel Walter
Olaf Meier
Ralph Ammer
Raúl Krauthausen
Sandra Mamitzsch
Sandrine Landrix
Sarah Sheik
Sascha Kaus
Shawn Thomson
Shermin Voshmgir
Tobias Laukemper
Tomàš Sedláček
Vinay Gupta
Yasmine Orth

Printed in Great Britain
by Amazon